# SpringerBriefs in Electrical and Computer Engineering

More information about this series at http://www.springer.com/series/10059

**SpringerBriefs in Electrical and Computer Engineering**

More information about this series at http://www.springer.com/series/10059

Amila Tharaperiya Gamage
Xuemin (Sherman) Shen

# Resource Management for Heterogeneous Wireless Networks

 Springer

Amila Tharaperiya Gamage
CISCO
Kanata, ON, Canada

Xuemin (Sherman) Shen
Department of Electrical and Computer
  Engineering
University of Waterloo
Waterloo, ON, Canada

ISSN 2191-8112          ISSN 2191-8120   (electronic)
SpringerBriefs in Electrical and Computer Engineering
ISBN 978-3-319-64267-3          ISBN 978-3-319-64268-0   (eBook)
DOI 10.1007/978-3-319-64268-0

Library of Congress Control Number: 2017947515

Printed on acid-free paper

This Springer imprint is published by Springer Nature
The registered company is Springer International Publishing AG
The registered company address is: Gewerbestrasse 11, 6330 Cham, Switzerland

# Preface

*Resource Management for Heterogeneous Wireless Networks* was written to provide an in-depth discussion on how to efficiently manage resources of heterogeneous wireless networks and how to design resource allocation algorithms to suit real-world conditions. Efficiently managing resources of the networks is more crucial now, than ever before, to meet users' rapidly increasing demand for higher data rates, better quality of service (QoS), and seamless coverage. Designing resource allocation algorithms to suit real-world conditions is also important, as the algorithms should be deployable and perform well in real networks. For example, two of the conditions considered in this book are resource allocation intervals of different networks are different and small cell base stations have limited computational capacity. To address the first condition, resource allocation algorithms for interworking systems are designed to allocate resources of different networks at different timescales. To address the second condition, resource allocation algorithms are designed to be able to run at cloud computing servers. More of such conditions, algorithms designed to suit these conditions, modeling techniques for various networks, and performance analysis of the algorithms are discussed in the book. These discussions will motivate new ideas in the research community on how to efficiently manage heterogeneous wireless network resources and will also be highly valuable to the people who plan, manage, and optimize the networks.

The book is organized as follows. Chapter 1 presents the techniques that can be incorporated within heterogeneous wireless networks to achieve higher data rates, better QoS, and seamless coverage. These techniques include interworking of the networks, user multi-homing, and device-to-device (D2D) communication. The advantages and disadvantages of these techniques are also discussed. Chapter 2 provides an overview of how and where each of these techniques can be applied in a typical heterogeneous wireless network. Chapters 3, 4, and 5 focus on three key areas of the network: (1) an area with interworking cellular networks and wireless local area networks (WLANs), (2) an area with interworking and D2D communication underlaying cellular networks and WLANs, and (3) an area with interworking macro cell and hyper-dense small cell networks. Each of these three chapters provides a comprehensive literature review on the existing resource allocation schemes, discusses challenges to allocate resources and the potential solutions, and derives resource allocation algorithm based on these solutions. Chapter 6 summarizes the conclusions and discusses future directions.

We would like to thank Prof. Jon W. Mark, Prof. Weihua Zhuang, Dr. Hao Liang, Dr. Ran Zhang, Dr. Qinghua Shen, Dr. Muhammad Ismail, Dr. Ning Zhang, Dr. Shamsul Alam, and the members of BroadBand Communications Research (BBCR) group at the University of Waterloo for their valuable comments and suggestions during the research work and preparation of the book.

Kanata, ON, Canada                                    Amila Tharaperiya Gamage
Waterloo, ON, Canada                                  Xuemin (Sherman) Shen

# Contents

# List of Abbreviations

| | |
|---|---|
| **3GPP** | 3rd Generation Partnership Project |
| **AAA** | Authorization, Authentication and Accounting |
| **ABS** | Almost blank subframes |
| **A-GNSS** | Assisted Global Navigation Satellite Systems |
| **AP** | Access point |
| **ASN** | Access service network |
| **BM1** | Benchmark resource allocation algorithm-1 |
| **BM2** | Benchmark resource allocation algorithm-2 |
| **BS** | Base station |
| **CCI** | Co-channel interference |
| **CCS** | Centralized control server |
| **CDMA** | Code division multiple access |
| **CFP** | Contention-free period |
| **CoMP** | Coordinated multipoint |
| **CP** | Contention period |
| **CSI** | Channel state information |
| **CTS** | Clear to send |
| **D2D** | Device-to-device |
| **DCF** | Distributed coordination function |
| **DL** | Downlink |
| **DSL** | Digital subscriber line |
| **eNB** | Enhanced NodeB |
| **EPC** | Evolved packet core |
| **ePDG** | Evolved Packet Data Gateway |
| **FDD** | Frequency-division duplexing |
| **FFR** | Fractional frequency reuse |
| **FSMC** | Finite-state Markov channel |
| **HCF** | Hybrid coordination function |
| **HM** | Heuristic resource allocation algorithm |
| **ICI** | Intercarrier interference |

| | |
|---|---|
| **IEEE** | Institute of Electrical and Electronics Engineers |
| **IP** | Internet Protocol |
| **ISP** | Internet service provider |
| **LPP** | LTE Positioning Protocol |
| **LTE** | Long-term Evolution |
| **LTE-A** | LTE Advanced |
| **MAC** | Medium access control layer |
| **MIMO** | Multiple-input and Multiple-output |
| **MM** | Optimal MMDP-based resource allocation algorithm |
| **MMDP** | Multiple timescale Markov decision process |
| **MMSE** | Minimum mean square error |
| **MRC** | Maximal ratio combining |
| **NTPv4** | Network Time Protocol version 4 |
| **OFDM** | Orthogonal frequency-division multiplexing |
| **OFDMA** | Orthogonal frequency-division multiple access |
| **OTDOA** | Observed time difference of arrival |
| **PDG** | Packet Data Gateway |
| **PDN** | Packet Data Network |
| **PDN-GW** | PDN Gateway |
| **PHY** | Physical layer |
| **QoS** | Quality of service |
| **RB** | Resource block |
| **RTS** | Request to send |
| **SDT** | Sum of discounted throughputs |
| **S-GW** | Serving Gateway |
| **SI** | Satisfaction Index |
| **SINR** | Signal-to-interference-plus-noise ratio |
| **TDD** | Time-division duplexing |
| **TXOP** | Transmission Opportunity |
| **UE** | Users' equipment |
| **UL** | Uplink |
| **WAG** | WLAN Access Gateway |
| **WiMAX** | Worldwide Interoperability for Microwave Access |
| **WLAN** | Wireless local area network |

# List of Symbols

| | |
|---|---|
| $A_{u,l}^L$ | Resource allocation decisions made at lower level at the beginning of the $(u, l)$th time slot |
| $A_u^U$ | Resource allocation decisions made at upper level at the beginning of the $u$th time slot |
| $B$ | Bandwidth of a network |
| $B^W$ | Bandwidth of WLAN |
| $C$ | Number of clusters in the network |
| $D$ | Maximum allowed packet size in contention-based channel access |
| $d_l$ | The distance within which D2D users are allocated orthogonal cellular network resources |
| $d_p$ | Processing time at the cloud |
| $d_q$ | Queueing delay at the cloud |
| $d_t$ | Transmission delay when accessing cloud |
| $d_{Total}$ | Total delay when accessing cloud |
| $\mathbb{D}^L$ | Lower-level policy |
| $\mathbb{D}^U$ | Upper-level policy |
| $E_1(\cdot)$ | Exponential integral |
| $F^s$ | Linear polyhedron at $s$th iteration |
| $h$ | Channel gain |
| $\tilde{h}_t$ | Normalized complex channel gain at $t$th time slot |
| $H_{ukt}$ | $|\tilde{h}_t|^2$ of $u$th user over $k$th subcarrier |
| $H_{vkt}^u$ | Normalized power gain of the channel over $k$th subcarrier between $v$th user and the BS to which $u$th user is connected to, during $t$th fast timescale time slot |
| $I_c$ | Maximum CCI received by the eNB |
| $I_{uk}$ | Normalized average interference to $u$th user over $k$th subcarrier |
| $I_{ukt}$ | Normalized interference to $u$th user over $k$th subcarrier during $t$th fast timescale time slot |
| $\mathcal{K}$ | Set of subcarriers available for the entire network |
| $\mathcal{K}^C$ | Set of OFDM subcarriers available in cellular network |

| | |
|---|---|
| $\mathcal{K}^{CF}$ | Set of contention-free TXOPs available in WLAN |
| $K_S$ | Number of possible states for each channel |
| $L$ | Number of fast timescale time slots within one slow time-scale time slot |
| $\mathbf{L}$ | A vector consists of $L_i$'s |
| $L_0$ | Number of fast timescale time slots in $T_0$ |
| $L_i$ | Duration of a data packet transmitted by the $i$th user |
| $L^U(\cdot)$ | Lagrangian for the problem $\mathcal{P}2$ |
| $L^{U(2)}(\cdot)$ | Lagrangian for the problem $\mathcal{P}3$ |
| $\mathcal{M}^{(c)}$ | Set of macro cells in $c$th cluster |
| $n$ | Total noise plus interference power |
| $N$ | Number of users in the interworking system |
| $N_0$ | Single-sided power spectral density of additive white Gaussian noise |
| $N_W$ | Number of users allocated for the contention-based channel access of WLAN |
| $\mathcal{N}^{(c)}$ | Set of small cells in $c$th cluster |
| $p_{ukt}$ | Transmit power of $u$th user during $t$th fast timescale time slot over $k$th subcarrier |
| $\mathbf{P}$ | A vector consists of $P_{uk}$'s |
| $P_{T,i}$ | Total average power available for the $i$th user |
| $P_{uk}$ | Average transmit power of $u$th user over $k$th subcarrier during next slow timescale time slot |
| $P_i^C$ | Total transmit power used by $i$th user for communications over cellular network |
| $P_{i,k}^C$ | Transmit power level of $i$th user over $k$th subcarrier |
| $P_{avg,i}^C$ | Average power usage of $i$th user for cellular network during one time slot in slow timescale |
| $P_{tot,i,l}^C(\cdot)$ | Total power allocated by the $i$th user for cellular network during the $(u, l)$th time slot |
| $\mathbf{P}^{CB}$ | A vector which consists of transmit power levels of users in $\mathcal{S}^{CB}$ during contention-based channel access |
| $\mathbf{P}_{-1}^{CB}$ | A vector which consists of transmit power levels of users in $\mathcal{S}^{CB}$ except the $i$th user, during contention-based channel access |
| $P_i^{CB}$ | Transmit power level of $i$th user during contention-based channel access |
| $P_{avg,i}^{CB}(\cdot)$ | Average power usage of $i$th user for contention-based channel access during one time slot in slow timescale |
| $P_{i,j}^{CF}$ | Transmit power level of $i$th user over $j$th contention-free TXOP |
| $P_{\psi_0^U \psi_1^U}^U$ | Probability of upper-level state change from $\psi_0^U$ to $\psi_1^U$ |
| $P_{\psi_{0,0}^L \psi_{1,0}^L}^L$ | Probability of lower-level state change from $\psi_{0,0}^L$ to $\psi_{1,0}^L$ |
| $P_{\psi_{u,0}^L \psi_{u,1}^L}^{L(2)}$ | Probability of lower-level state change from $\psi_{u,0}^L$ to $\psi_{u,1}^L$ |
| $P_i^W$ | Total transmit power used by $i$th user for communications over WLAN |
| $r_{ukt}$ | Throughput achieved by $u$th user during $t$th fast timescale time slot over $k$th subcarrier |

| | |
|---|---|
| $r_{i,u,l}^{L}(\cdot)$ | Throughput achieved by the $i$th user at the lower level during $(u, l)$th time slot in fast timescale |
| $r_{i,u}^{U}(\cdot)$ | Throughput achieved by the $i$th user at the upper level during $u$th time slot in slow timescale |
| $R_{Dmin,i}$ | Minimum data rate required for data traffic services of $i$th user |
| $R_{Vmin,i}$ | Minimum data rate required for voice traffic services of $i$th user |
| $R_{min}$ | Minimum required data rate |
| $R_{uk}$ | Average throughput of $u$th user over $k$th subcarrier during next slow timescale time slot |
| $R_i^{C}$ | Throughput achieved by $i$th user via cellular network |
| $R_{i,k}^{C}(\cdot)$ | Throughput achieved by $i$th user over $k$th subcarrier |
| $R_i^{CB}(\cdot)$ | Average throughput achieved by $i$th user via contention-based channel access |
| $R_i^{C(D)}$ | Throughput achieved by $i$th user via cellular network using D2D mode |
| $R_{i,j}^{CF}(\cdot)$ | Throughput achieved by $i$th user over $j$th contention-free TXOP |
| $R_i^{C(T)}$ | Throughput achieved by $i$th user via cellular network using traditional mode |
| $R_{min,i}^{D2D}$ | Minimum data rate required for $i$th user's D2D communications |
| $R_{i,u}^{L}(\cdot)$ | SDT achieved by the $i$th user at the lower level over the $u$th time slot in slow timescale |
| $R_{min,i}^{ND}$ | Minimum data rate required for $i$th user's non-D2D communications |
| $R_i^{U}(\cdot)$ | SDT achieved by the $i$th user at the upper level |
| $R_i^{W}$ | Throughput achieved by $i$th user via WLAN |
| $R_i^{W(D)}$ | Throughput achieved by $i$th user via WLAN using D2D mode |
| $R_i^{W(T)}$ | Throughput achieved by $i$th user via WLAN using traditional mode |
| $\mathcal{S}_1$ | Set of users who are allocated resource during the first step |
| $\mathcal{S}_2$ | Set of users who are allocated resource during the second step |
| $\mathcal{S}_M$ | Set of low-mobility users within WLAN coverage |
| $\mathcal{S}_N$ | Set of all users |
| $\mathcal{S}_S$ | Set of all users except the users in $\mathcal{S}_M$ |
| $\mathcal{S}^{CB}$ | Set of users communicating through contention-based channel access |
| $\mathcal{S}^{CF}$ | Set of users communicating through contention-free channel access |
| $SI_D$ | Satisfaction index for data traffic |
| $SI_V$ | Satisfaction index for voice traffic |
| $SI_{x,y}$ | Satisfaction index achieved by $y$ algorithm for $x$ traffic class |
| $T_0$ | Duration from a beginning of a time slot to the point at which the BSs send CSI to the cloud |
| $T_{ACK}$ | Duration of an acknowledgment in contention-based channel access |
| $T_{AIFS}$ | Duration of arbitration interframe space in contention-based channel access |
| $T_{coh}$ | Coherence time of a wireless channel |
| $T_{CF}$ | Durations of a contention-free TXOP |
| $T_{CFP}$ | Durations of CFP |

| | |
|---|---|
| $T_{CP}$ | Durations of CP |
| $T_{CTS}$ | Duration of CTS message in contention-based channel access |
| $T_F$ | Duration of a fast timescale time slot |
| $T_P$ | CFP repetition period |
| $T_{RTS}$ | Duration of RTS message in contention-based channel access |
| $T_S$ | Duration of a slow timescale time slot |
| $T_{SIFS}$ | Duration of short interframe space in contention-based channel access |
| $T^L$ | Duration of a time slot in fast timescale |
| $T^U$ | Duration of a time slot in slow timescale |
| $\mathcal{U}^{(c)}$ | Set of users in $c$th cluster |
| $\mathcal{U}_b^{(c)}$ | Set of users connected to $b$th cell in $c$th cluster |
| $V_L$ | Number of fast time-scale time slots within a slow timescale time slot |
| $V^s$ | Set of vertices of the linear polyhedron $F^s$ |
| $x_{uk}$ | Auxiliary variables |
| $\mathbf{x}$ | A vector consists of $x_{uk}$'s |
| $Y$ | Percentage of users who can communicate using D2D mode |
| $\alpha_{i,k}^C$ | SINR of the channel between $i$th user and cellular BS over $k$th subcarrier with unit transmit power |
| $\alpha_i^W$ | SINR of the channel between $i$th user and WLAN AP with unit transmit power |
| $\beta$ | Discount factor for the lower level |
| $\boldsymbol{\gamma}$ | Vector of dual variables corresponding to contention-free TXOP allocation constraints |
| $\gamma_j$ | Dual variable corresponds to $j$th contention-free TXOP allocation constraint |
| $\Delta f$ | Bandwidth of an OFDM subcarrier |
| $\theta$ | Discount factor for the upper level |
| $\boldsymbol{\lambda}$ | Vector of dual variables corresponding to data traffic constraints |
| $\lambda_i$ | Dual variable corresponds to data traffic constraint of $i$th user |
| $\boldsymbol{\mu}$ | Vector of dual variables corresponding to total power constraints |
| $\mu_i$ | Dual variable corresponds to total power constraint of $i$th user |
| $\mu_{uk}$ | Water level for $u$th user over $k$th subcarrier |
| $\boldsymbol{\xi}$ | Vector of dual variables corresponding to voice traffic constraints |
| $\xi_i$ | Dual variable corresponds to voice traffic constraint of $i$th user |
| $\rho$ | Correlation coefficient |
| $\rho_{i,k}^C$ | $\rho_{i,k}^C = 1$ if $k$th subcarrier is allocated for $i$th user or $\rho_{i,k}^C = 0$ otherwise |
| $\rho_{i,j}^{CF}$ | $\rho_{i,j}^{CF} = 1$ if $j$th contention-free TXOP is allocated for $i$th ($i \in \mathcal{S}_M$) user or $\rho_{i,j}^{CF} = 0$ otherwise |
| $\sigma_0$ | Duration of an empty slot in contention-based channel access |
| $\sigma^2$ | Average power gain of a channel divided by the noise power |
| $\sigma_{uk}^2$ | Average normalized power gain of the channel over $k$th subcarrier between $u$th user and the BS to which $u$th user is connected to |
| $(\sigma_{vk}^u)^2$ | Average normalized power gain of the channel over $k$th subcarrier between $v$th user and the BS to which $u$th user is connected to |

| | |
|---|---|
| $\tau$ | Probability of a user transmits a packet in a randomly chosen time slot during contention-based channel access |
| $\psi_{u,l}$ | System state, i.e., $\{\psi_u^U, \psi_{u,l}^L\}$ |
| $\psi_{u,l}^L$ | State of the lower level during $(u, l)$th time slot in fast timescale |
| $\psi_u^U$ | State of the upper level during $u$th time slot in slow timescale |
| $\Psi^L$ | Set of all the possible states of lower level |
| $\Psi^U$ | Set of all the possible states of upper level |
| $\tilde{w}$, $\tilde{w}_t$ | Complex Gaussian random variables |
| $\Omega$ | Average of square channel gain |

# Chapter 1
# Introduction

The volume of global mobile data traffic that wireless networks should support by year 2021 is expected to be as high as seven times of its value in year 2016 [1]. This increase is due to two main reasons: (1) recent advancements in mobile industry which has dramatically increased the number of smart mobile devices, such as smart phones, tablets and PDAs, operating in any geographical region; and (2) increase in the number of data hungry applications that run on these devices, such as video streaming, YouTube, Google Maps and Facebook. In addition to high volume of data, wireless networks are also expected to provide seamless service coverage and support various applications' diverse quality-of-service (QoS) requirements.

To achieve future capacity, coverage and QoS demands, three key techniques can be used: (1) add more spectrum to the networks; (2) increase the spectrum efficiency; and (3) add more small cells to the networks. Adding more spectrum to the networks may not be a viable solution as most of the underutilized spectrum lies in very high frequency ranges, which are not cellular communication-friendly due to their propagation characteristics limit the communication distance. For example, millimeter-wave signals are significantly attenuated even by very thin obstacles, such as narrow walls and rain [2, 3]. Thus, underutilized millimeter-wave frequency band between 30 and 300 GHz is suitable only for indoor communications and small cells.

Efficient utilization of the spectrum available at lower frequency ranges, such as 800, 1.7 and 800 MHz, 1.7 GHz and 2.6 GHz communications bands, is essential to facilitate robust outdoor long range communications. Promising techniques to enhance the spectrum efficiency are multiple-input and multiple-output (MIMO) communication techniques, frequency reuse, interworking of heterogeneous networks, and device-to-device (D2D) communication. MIMO techniques, such as coordinated multi-point (CoMP) and beamforming, increase the spectrum efficiency and provide robust communication by exploiting spatial diversity [4]. Frequency reuse improves the spectrum efficiency by allowing multiple cells to reuse the same frequency to increase the number of bits transmitted per frequency [5, 6].

© The Author(s) 2018                                                                                  1
A.T. Gamage, X.(S.) Shen, *Resource Management for Heterogeneous Wireless
Networks*, SpringerBriefs in Electrical and Computer Engineering,
DOI 10.1007/978-3-319-64268-0_1

Interworking enhances the spectrum efficiency by efficiently utilizing resources available at multiple networks. First it enables user multi-homing which allows users to simultaneously communicate over multiple networks. Then it jointly allocates resources (e.g., frequency and transmit power) available at multiple networks in a user QoS requirements satisfying manner. Moreover, interworking provides users with seamless network coverage by merging coverage of individual networks. D2D communication result in a high spectrum efficiency as it allows direct communication among the users in proximity [7]. Direct communication among these users result in high throughputs due to short communication distances, and saves network resources as it uses single hop communication. Traditional communication uses two hops; first hop is between message transmitting user and base station (BS), and second hop is between BS and message receiving user.

Small cells increase wireless network capacity and QoS by bringing network closer to the users, as it results in high user throughputs due to strong wireless channels [8]. Furthermore, small cells increase frequency reuse as they are low-powered small-radius cells and a large number of them are deployed in a network.

Interworking of heterogeneous networks is one of the most promising techniques that can provide better network capacity, coverage and QoS, as it can leverage multiple networks available at high volumes of mobile traffic demanding areas, such as, office buildings, airports and hotspots. These areas are typically covered by cellular networks and wireless local area networks (WLANs). However, interworking cannot provide such network performance enhancements when there is coverage of only one network, for example, at rural areas and cell edges. To provide better network performance in these areas while achieving network-wide uniform performance improvements, interworking can be integrated with D2D communication, frequency reuse, and small cell deployment techniques.

Rest of the chapter discusses advantages of and challenges for interworking, integration of D2D communication with interworking, interworking of macro cell and highly dense small cell networks, and contributions of the book.

## 1.1   Interworking of Wireless Networks

Wireless networks with diverse radio access technologies can be interconnected via interworking mechanisms in order to provide users with higher throughputs, better network coverage, and better QoS support [9].

Seamless network coverage is provided by merging coverages of individual networks, and it is achieved by allowing users to access their desired services utilizing resources available in multiple networks. Figure 1.1 demonstrates coverage extension achieved in a cellular/WLAN interworking system. In this example, user-A initiates a call using the cellular network, and the call is continued via WLAN without any interruption as user-A moves away from the cellular BS.

Interworking enhances user throughput and QoS by jointly allocating resources of multiple networks and enabling user multi-homing. Jointly allocating resources

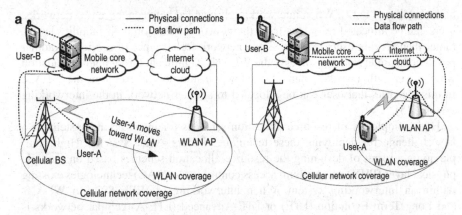

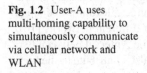

**Fig. 1.1** Coverage extension achieved by cellular/WLAN interworking. (**a**) User-A moves while calling user-B. (**b**) When the cellular signal strength becomes weak, user-A continues the call via WLAN

**Fig. 1.2** User-A uses multi-homing capability to simultaneously communicate via cellular network and WLAN

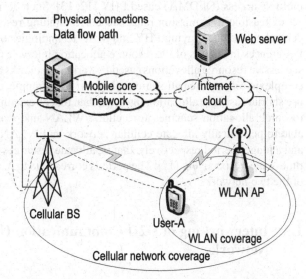

of multiple networks among all the users in the interworking system increases the efficiency of resource utilization, compared to individually allocating available resources in each network. With multi-homing enabled, users' equipment (UEs) with multiple radio interfaces are able to simultaneously communicate over multiple networks as shown in Fig. 1.2. Since UEs can access services simultaneously utilizing resources from multiple networks in an optimal manner, multi-homing further improves the efficiency of resource utilization in the interworking system [10].

Interworking can overcome the shortcomings of individual networks. For example, WLANs provide services at much lower charges with limited support for mobility while cellular networks provide services at higher charges with support for high mobility [11–14]. When interworking is enabled between these two networks, users can be allocated resources from the networks based on their mobility levels, rather than forcing users to use a single network for all types of services. Another problem that interworking solves is the WLANs' significant throughput degradation when a user with a weaker channel or a low transmit power level joins the WLAN. In this scenario, that user can be allocated to another network in the interworking system [15].

High complexity of resource allocation schemes for interworking systems is a key challenge for deploying these techniques in large scale systems. High complexity is a result of designing the resource allocation schemes based on multiple physical layer (PHY) and medium access control layer (MAC) technologies existing within an interworking system. When interworking of IEEE 802.11n WLANs and Long Term Evolution (LTE) or LTE-Advanced (LTE-A) cellular networks is considered, WLAN uses a hybrid coordination function (HCF) based MAC and an orthogonal frequency division multiplexing (OFDM) based PHY, whereas cellular network uses a centrally coordinated MAC and a orthogonal frequency division multiple access (OFDMA) based PHY [12, 13]. Such designing is necessary as the set of feasible transmission rates and corresponding resource allocation decisions depend on the underlaying PHY and MAC technologies of different networks [16]. Differences in lengths of the resource allocation intervals (i.e., interval between two successive resource allocations) used by various networks also contribute to the high complexity of resource allocation schemes. For example, existing cellular networks use shorter resource allocation intervals than that of existing WLANs [12, 13]. Thus, resource allocation schemes for cellular/WLAN interworking systems should be able to periodically allocate cellular network and WLAN resources with a shorter and a longer period respectively, that is, allocate resources over a faster and a slower time-scale, respectively [17]. The periods correspond to resource allocation intervals of the two networks.

## 1.2   Interworking of D2D Communication Underlaying Networks

Using device-to-device communication, users in proximity can directly communicate among themselves, without having to send data through a BS. Such direct communication is referred to as D2D communication mode. When communications from source to destination are routed through a BS, it is referred to as traditional communication mode. In Fig. 1.3, $UE_3$ and $UE_4$ communicate using D2D communication mode while $UE_1$ and $UE_2$ communicate using traditional communication mode. Furthermore, users can initiate D2D communication among themselves without any involvement from the network operators, or the network operators can initiate and control D2D communication between users. The latter is referred to as network assisted D2D communication mode, and it provides security and mobility support for users while reducing the interference caused by non-synchronized users [18].

**Fig. 1.3** D2D and traditional communications in a cellular network

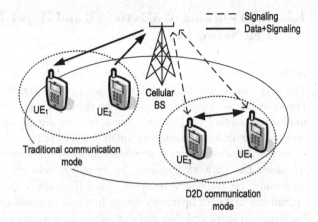

Enabling D2D communication in an interworking system provides two key benefits. First, D2D communication can be utilized to improve network performance at the areas where interworking cannot. Such areas include areas with coverage of only one network. D2D communication improves network performance in these areas as it allows direct communication between users in proximity, and incorporates reuse and hop gains to the network [18–22]. Reuse gain is achieved by simultaneously using the same set of resources for both D2D and traditional communication links [18, 19, 23]. Hop gain is a result of D2D communication links using either uplink (UL) or downlink (DL) resources only. Thus, by integrating D2D communication with interworking, more uniform performance improvements throughout the networks can be achieved. Second, D2D communication reduces the cost of service as data traffic is offloaded from mobile network to the D2D links [24, 25]. Furthermore, D2D communication can be enabled without deploying any extra hardware.

Two main types of resource allocation mechanisms are available for D2D communication links: (1) from a fixed set of resources, each D2D communication link is allocated resources; and (2) resources for D2D and traditional communication links are jointly allocated. Each of these two resource allocation mechanisms can be categorized into two types: orthogonal and non-orthogonal resource sharing. With orthogonal resource sharing, resource sets allocated for D2D and traditional communication links are non-overlapping. With non-orthogonal resource sharing, D2D and traditional communication links share the same resource set. Non-orthogonal resource sharing is more efficient as all D2D and traditional communication links are free to utilize all the available resources. However, it causes co-channel interference (CCI) among D2D and traditional communication links.

Several technical challenges make the resource allocation for D2D communication underlaying interworking systems complicated: (1) involvement of multiple PHY and MAC technologies of different networks [10, 12, 13], (2) for each user over each network, either traditional or D2D communication mode should be selected such that hop and reuse gains are maximized, and (3) interference management [18].

## 1.3  Interworking of Macro Cell and Hyper-Dense Small Cell Networks

High volumes of mobile data traffic in areas, such as busy streets, city centers and shopping malls, can be served by hyper densely deployed small cells. Network operators as well as subscribers can deploy these cells in both planned and unplanned manner [26]. Fiber-optic cables or digital subscriber lines (DSL) can provide backhaul connectivity to small cell BSs [27].

Hyper-dense small cell deployments provide three main benefits. First, it increases the network capacity by bringing network closer to the users and increasing the frequency reuse [8, 28, 29]. Second, it reduces deployment and operational costs for operators due to low cost of small cell BSs, low cost of DSL backhauls to small cell BSs and that subscribers install and operate majority of the small cell BSs [27]. Third, small cells can use the under utilized higher frequency bands due to small targeted cell coverage areas.

By enabling interworking between hyper-dense small cell and macro cell networks, better coverage, capacity and QoS support can be achieved. In hyper-dense small cell networks, there may be coverage holes due to unplanned deployment of BSs with small coverage areas. Moreover, due to the same reasons, calls may be dropped when users are highly mobile. These shortcomings are eliminated when interworking between macro cell and hyper-dense small cell networks is enabled. In this interworking system, coverage holes of the small cell network are covered by the macro cell network, and the highly mobile users are also served by the macro cell network. An example of such interworking system is shown in Fig. 1.4.

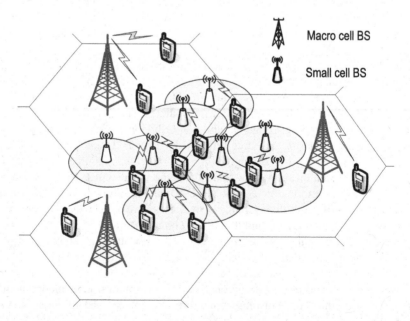

**Fig. 1.4** Interworking of macro cell and hyper-dense small cell networks

Four challenges make the resource allocation for interworking macro cell and hyper-dense small cell networks complicated. First, severe CCI exists in this system due to short distances among densely deployed small cells [8, 30]. Thus, resources of all the small cells in proximity have to be jointly allocated to ensure a tolerable CCI level. Second, network load considerably changes and moves across the network with time. For example, during lunch time the load will be high near restaurants while in the afternoon the load will be high near coffee shops. Such network load variations should be considered in order to optimize available spectrum utilization. Third, limited backhaul capacities of the small cell BSs should be considered in the resource allocation problem to prevent bottlenecking the backhauls as a result of high volume of user data and control signaling. Fourth, small cell BSs cannot run complex resource allocation algorithms due to limited computational capacity available at these low-cost BSs [31].

## 1.4 Contributions and Outline of the Book

Interworking of heterogeneous wireless networks is a promising technique that can enhance network capacity, coverage and QoS to meet future service demands. Interworking is able to provide such enhancements to the network as it leverages multiple networks available at most of the high capacity demanding areas. Furthermore, interworking can be integrated with D2D communication, frequency reuse, and small cell deployment techniques to achieve network-wide uniform capacity, coverage and QoS improvements. A key problem that needs to be solved before these techniques can be applied in practical networks is the efficient resource, such as frequency and transmit power, allocation when these techniques are applied. Therefore, this book describes appropriate resource allocation frameworks as well as optimal and sub-optimal resource allocation algorithms for interworking heterogeneous wireless networks which also integrate D2D communication and small cell deployment techniques.

The remainder of the book is organized as follows.

Chapter 2 presents the heterogeneous wireless network model.

Chapter 3 describes the resource allocation framework for cellular/WLAN interworking system. Then it develops an optimal multiple time-scale Markov decision process (MMDP) based and a low-complex heuristic resource allocation schemes for the system. These schemes are designed considering multi-homing capable users with voice and data traffic requirements, and underlying PHY and MAC layer technologies of the networks.

Chapter 4 presents the technical challenges for selecting communication modes and allocating resources in a D2D communication underlaying cellular/WLAN interworking system. Next, a semi-distributed resource allocation scheme, which addresses these challenges, is presented. Related implementation issues are also investigated.

Chapter 5 describes resource allocation for interworking macro cell and hyper-dense small cell networks, with assistance of cloud computing. In the resource allocation algorithm presented in this chapter, the complex part of the algorithm runs at cloud computing servers, as low-cost small cell BSs are not able to run complex algorithms. The challenges due to use of cloud computing is also investigated.

Chapter 6 presents the conclusions and the future research directions.

# Chapter 2
# Heterogeneous Wireless Networks

This chapter first presents a heterogeneous wireless network which consists of WLANs and cellular macro cell and hyper-dense small cell networks. Second, it describes the scenarios where D2D communication and interworking among these networks can be applied. Each of these scenarios are investigated in detail in the following chapters. Third, achievable user throughputs over each network and user power consumptions are derived based on MAC and PHY technologies of the networks.

## 2.1 Network Overview

A heterogeneous wireless network that uses interworking, D2D communication and small cell deployment techniques to enhance network throughput, coverage and QoS is shown in Fig. 2.1. This network consists of WLANs and cellular macro cell and hyper-dense small cell networks. WLANs are similar to IEEE 802.11n WLANs [12] while macro cell and hyper-dense small cell networks are similar to LTE and LTE-A networks [13], in terms of their PHY and MAC technologies. In this network, there are three types of areas of interest: (1) areas which are covered by the macro cell network only, (2) areas which are covered by both macro cell network and a WLAN, and (3) areas which are covered by both macro cell and hyper-dense small cell networks. Users in the first type areas communicate through the macro cell network only. In Fig. 2.1, $UE_3-UE_5$, $UE_9$ and $UE_{10}$ are in such areas. In the second type areas, interworking has been enabled among macro cell network and WLANs. Therefore, users in these areas are able communicate through macro cell network or WLAN, or through both networks using UE multi-homing capability. For example, $UE_2$ communicates simultaneously using both WLAN and the macro cell network. Using D2D communication, users in proximity can directly communicate among themselves, for example, $UE_4$ and $UE_5$, and $UE_{11}$ and $UE_{12}$. As $UE_{11}$ and $UE_{12}$ are

© The Author(s) 2018
9
A.T. Gamage, X.(S.) Shen, *Resource Management for Heterogeneous Wireless Networks*, SpringerBriefs in Electrical and Computer Engineering, DOI 10.1007/978-3-319-64268-0_2

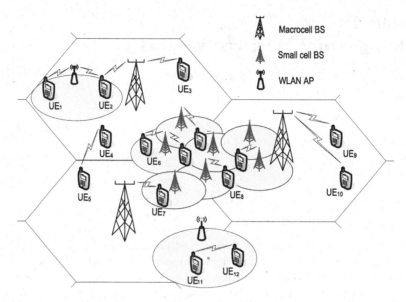

**Fig. 2.1**  A heterogeneous wireless network

in coverage areas of macro cell network and a WLAN, they can form the D2D link using either network or both networks. In the third type areas, interworking has been enabled between macro cell and hyper-dense small cell networks. Therefore, users in this area can communicate using either network or both networks. For example, $UE_6$ and $UE_7$ communicate using the hyper-dense small cell network and the macro cell network, respectively.

## 2.2   Cellular Networks

Macro cell and hyper-dense small cell networks consist of centrally coordinated MAC layers and OFDMA based PHY layers. MAC layers centrally coordinate allocation of the network resources, such as OFDMA subcarriers and transmit power of BSs and UEs. Therefore, user throughputs via cellular networks can be calculated as follows.

The total frequency band of a network is divided into OFDMA subcarriers with bandwidth of $\Delta f$. Over the $k$th subcarrier, the $i$th user's maximum achievable error free data rate is

$$R_{i,k}^C(P_{i,k}^C) = \Delta f \log_2(1 + \alpha_{i,k}^C P_{i,k}^C), \tag{2.1}$$

where $\alpha_{i,k}^C$ is the signal-to-interference plus noise ratio (SINR) of the channel between the $i$th user and the cellular BS over the $k$th subcarrier with unit transmitted power; and $P_{i,k}^C$ is the transmit power of the $i$th user over the $k$th subcarrier.

## 2.3 WLANs

In WLANs, PHY layer is OFDM based while MAC grants transmission opportunities (TXOP) to users using two channel access mechanisms: (1) contention-free polling based channel access during contention-free period (CFP); and (2) contention-based channel access during contention period (CP). Operation of these channel access mechanisms is shown in Fig. 2.2. CFP and CP alternate over time and they repeat once every $T_P$. Their durations are denoted by $T_{CFP}$ and $T_{CP}$, respectively. In contention-free polling based channel access, a centralized polling mechanism grants TXOPs to users. Each of these TXOPs allows a user to transmit his data over a fixed time duration defined by $T_{CF}$. In contention-based channel access, users contend for the channel to obtain TXOPs. When a user obtain a TXOP, he can send a data packet of $D$ bits as channel capture is not allowed. In this channel access, to solve hidden terminal problem and to improve efficiency of the WLAN by reducing the duration of collisions, four-way handshaking scheme with request-to-send (RTS) and clear-to-send (CTS) messages is used. Contention-free channel access is more suitable for constant bit rate voice traffic while contention-based channel access is more suitable for variable bit rate data traffic [32].

### 2.3.1 Contention-Free Channel Access

In contention-free channel access, TXOPs are allocated by a centralized polling mechanism. Therefore, the $i$th user's maximum achievable error free data rate using the $j$th contention-free TXOP is

$$R_{i,j}^{CF}(P_{i,j}^{CF}) = B^W \log_2(1 + \alpha_i^W P_{i,j}^{CF}), \qquad (2.2)$$

where $\alpha_i^W$ is the SINR of the channel between the $i$th user and the WLAN access point (AP) with unit transmit power; $P_{i,j}^{CF}$ is the transmit power of the $i$th user over the $j$th contention-free TXOP; and $B^W$ is the bandwidth of the WLAN.

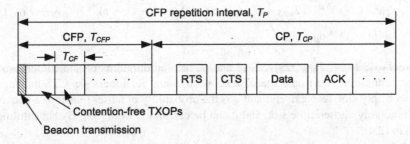

**Fig. 2.2** Operation of contention-free and contention-based channel access mechanisms

Wireless channels are assumed to remain unchanged within their coherence time. Thus, resource allocation interval for the WLANs is selected to be shorter than or equivalent to the channel coherence time, and $T_P$ is selected to be shorter than or equivalent to the selected resource allocation interval. Consequently, SINRs (i.e., $\alpha_i^W$) remain unchanged within a resource allocation interval, and also they are the same for contention-free and contention-based channel access mechanisms. Therefore, $\alpha_i^W$ does not have the subscript $j$ to denote the TXOP.

### 2.3.2  Contention-Based Channel Access

In contention-based channel access, users obtain TXOPs in a random manner. Moreover, during an obtained TXOP, data transmission may not be successful due to collisions of different users' data packets. Thus, during contention-based channel access period, average user throughputs and average power consumptions of the users are considered, and they can be calculated as follows.

Let $N_W$ denote the number of users who use contention-based channel access and $L_i(P_i^{CB})$ denote the duration of a successful packet transmission of the $i$th user who uses contention-based channel access. **L** denotes a vector which consists of $L_i(P_i^{CB}), i \in \{1, \ldots, N_W\}$. Then, $L_i(P_i^{CB})$ is given by

$$L_i(P_i^{CB}) = \frac{D}{B^W \log_2(1 + \alpha_i^W P_i^{CB})}, \tag{2.3}$$

where $P_i^{CB}$ is the transmit power of the $i$th user during the CP.

From [12, 33, 34], average throughput achieved by the $i$th user who uses contention-based channel access during a CP is then given by

$$R_i^{CB}(\mathbf{L}) = \frac{\tau(1-\tau)^{N_W-1}D}{T_1 + N_W\tau(1-\tau)^{N_W-1}\sum_{j=1}^{N_W} L_j(P_j^{CB})}, \tag{2.4}$$

where $T_1$ can be calculated as:

$$T_1 = N_W\tau(1-\tau)^{N_W-1}(T_{CTS} + T_{ACK} + 3T_{SIFS}) + (1 - (1-\tau)^{N_W})(T_{RTS} + T_{AIFS})$$
$$+ (1-\tau)^{N_W}\sigma_0,$$

where $T_{SIFS}$, $T_{AIFS}$, $T_{ACK}$, $T_{RTS}$, $T_{CTS}$ and $\sigma_0$ are the durations of short interframe space, arbitration interframe space, acknowledgment, RTS message, CTS message and an empty slot, respectively; and $\tau$ is the probability of a user transmits a packet in a randomly chosen time slot, and it can be calculated as in [33]. By substituting (2.3) to (2.4),

$$R_i^{CB}(\mathbf{P}^{CB}) = \frac{\tau(1-\tau)^{N_W-1}D}{T_1 + \dfrac{DN_W\tau(1-\tau)^{N_W-1}}{B^W}\sum_{j=1}^{N_w}\dfrac{1}{\log_2(1+P_j^{CB}\alpha_j^W)}}, \tag{2.5}$$

where $\mathbf{P}^{CB}$ is a vector consisting of $P_i^{CB}, i \in \{1,\ldots,N_W\}$. From (2.5), it can be seen that $R_i^{CB}(\mathbf{P}^{CB})$ is the same for all the users who use contention-based channel access. Furthermore, $R_i^{CB}(\mathbf{P}^{CB})$ is a concave function when $N_W$ is fixed, and it is proved in Appendix A.1.

Average transmit power of a user for contention-based channel access through the WLAN interface during a CP can be determined as follows. Since the probability of a successful transmission during a CP is $\tau(1-\tau)^{N_W-1}$, average transmit power of the $i$th user who uses contention-based channel access is given by

$$P_{avg,i}^{CB}(\mathbf{P}^{CB}) = \tau(1-\tau)^{N_W-1}\frac{L_iP_i^{CB}}{T_{avg}}\frac{T_{CP}}{T_P}, \tag{2.6}$$

where $T_{avg}$ is the average duration of channel occupancy due to an event of successful transmission, collision, or empty slot in which no user transmits, and $T_{avg}$ can be calculated as in [33]. By simplifying (2.6),

$$P_{avg,i}^{CB}(\mathbf{P}^{CB}) = \begin{cases} \dfrac{T_{CP}P_i^{CB}R_i^{CB}(\mathbf{P}^{CB})}{T_PB^W\log_2(1+\alpha_i^WP_i^{CB})}, & \text{if } P_i^{CB} > 0 \\ 0, & \text{otherwise.} \end{cases} \tag{2.7}$$

## 2.4  Summary

In this chapter, first an overview of a heterogeneous wireless network is presented. Second, to enhance the capacity, coverage and QoS of this network, integration of interworking and D2D communication and deploy hyper-dense small cell networks are discussed. Third, details of the PHY and MAC technologies of cellular networks and WLANs are presented. Analytical expressions of user throughputs and transmit power levels are also derived based on the underlying PHY and MAC technologies of the networks.

# Chapter 3
# Resource Allocation for Cellular/WLAN Interworking

This chapter investigates uplink resource allocation for cellular/WLAN interworking system in order to maximize the system throughput while satisfying QoS requirements of multi-homing users. The cellular network is similar to OFDMA based LTE/LTE-A networks. The WLAN is similar to hybrid coordination function (HCF) based IEEE 802.11n WLANs with both contention-based and contention-free polling-based channel access mechanisms. Resources to be allocated are user transmit power, subcarriers of the cellular network, transmission opportunities (TXOPs) via the two channel access mechanisms of the WLAN. A key challenge for allocating resources for this system is the high complexity of resource allocation algorithms. Reasons for such high complexity are the existence of multiple PHY and MAC technologies in the interworking system, and that the cellular network and the WLAN allocate resources at a fast and a slow time-scale.

This chapter first discusses challenges for resource allocation and related work. Second, the cellular/WLAN interworking system model is presented, and a resource allocation framework that operates on two time-scales is proposed. Third, based on the proposed resource allocation framework and PHY and MAC technologies of the two networks, a multiple time-scale Markov decision process (MMDP) based optimal resource allocation scheme and a low time complex heuristic resource allocation algorithm are derived. Fourth, performance of the proposed algorithms is evaluated and discussed.

## 3.1 Challenges for Resource Allocation

A key challenge for allocating resources for the cellular/WLAN interworking system is the high complexity of resource allocation algorithms due to two reasons: (1) existence of multiple PHY and MAC technologies within the system; and (2) the cellular network and the WLAN allocate resources at two different time-scales [17].

© The Author(s) 2018                                                                                    15
A.T. Gamage, X.(S.) Shen, *Resource Management for Heterogeneous Wireless Networks*, SpringerBriefs in Electrical and Computer Engineering, DOI 10.1007/978-3-319-64268-0_3

Existence of multiple PHY and MAC technologies makes the resource allocation algorithms complex, as these algorithms need to be designed considering multiple diverse PHY and MAC technologies. Such design consideration is needed to make feasible resource allocation decisions and determine accurate user throughputs [16, 34]. For example, bandwidth allocation decisions in an OFDMA based network should be in multiples of the subcarrier bandwidth, while user throughputs during the contention-based channel access of a WLAN should be calculated considering transmission collisions which occur due to the MAC scheme [12].

The cellular network and the WLAN allocating resources in two different time-scales makes the resource allocation algorithms complicated, where a time-scale defines how often a certain periodic event occurs and the start time of these events. These two networks allocate resources in two different time-scales as their resource allocation intervals are different. Resource allocation interval for a network is calculated based on the channel coherence time. Coherence time is much shorter in cellular networks than in WLANs, as cellular networks and WLANs are designed to support users with speeds up to $350\,\mathrm{km\,h^{-1}}$ and $3\,\mathrm{km\,h^{-1}}$, respectively [12, 13]. Thus, resource allocation interval of cellular networks is much shorter than that of WLANs. When resources of different networks in the system are allocated in different time-scales, predicting achievable throughputs and power consumptions over future time slots is required to optimally allocate resources. That is, when resources of the system are jointly allocated at a beginning of a slow time-scale time slot, it is necessary to predict achievable throughputs and power consumptions over future fast time-scale time slots that lie within the current slow time-scale time slot. As resource allocation algorithms have to make these predictions, they become complicated.

## 3.2   Related Work

Existing resource allocation schemes can be divided into three categories: (1) schemes using a single network interface of each UE at any given time [35–40]; (2) schemes using the multi-homing capability of UEs [10, 41–43]; and (3) schemes that are designed based on different PHY and MAC technologies [11, 34]. In [37], a load balancing scheme to improve resource utilization in cellular/WLAN interworking is presented. New voice and data calls are assigned to a network based on a set of precalculated probabilities. Assigned calls are re-distributed whenever necessary to another network by using dynamic vertical handoffs to reduce network congestion and improve QoS satisfaction. To further improve QoS satisfaction, in [35] the voice calls are allocated preferably for the cellular network. Resource allocation scheme proposed in [36] for WiMAX/WLAN interworking system assigns all streaming calls to the WiMAX network to guarantee QoS satisfaction; data calls that are served by the WiMAX network are preempted to free up bandwidth for the incoming streaming calls when required. The main advantage of these schemes in the first category is that they are easy to deploy, because each network uses its own resource allocation scheme and that designing a resource allocation scheme for an individual network is simpler than designing that for an interworking system.

Resource allocation schemes in the second category take advantage of the multi-homing capability of UEs. When UEs multi-home, UEs simultaneously access services through multiple networks. Therefore, flexibility to distribute resources of the interworking system among users increases. Thus, resource allocation schemes in the second category allocate resources of the system more efficiently than the schemes in the first category. In the second category schemes, for computational simplicity, it is typically assumed that the WLANs use a resource reservation protocol to avoid channel contention collisions. Hence, resources of the WLANs are modeled as frequency channels or time slots. Bandwidth allocation algorithms for UEs with different types of traffic requirements are proposed in literature. In [10], each network gives more priority to satisfy its own subscribers' QoS requirements, while utility fairness among users in the interworking system is maintained in [41]. A game theoretic approach for bandwidth allocation and admission control is investigated in [42]. Each network allocates its bandwidth for different service areas on a long-term basis based on the statistics of call arrivals; bandwidths for each service area from different networks are then allocated to users on a short-term basis. To ensure QoS satisfaction, a new call is accepted only if its minimum data rate requirement can be satisfied. Algorithms to allocate time slots in a WLAN and subcarriers in a cellular network subject to a proportional rate constraint are proposed in [43].

The third category includes resource allocation schemes proposed in [11, 34]. These schemes are designed based on PHY and MAC technologies of the different networks to ensure the feasibility of resource allocation decisions. For example, the effect of transmission collisions caused by the contention-based channel access in WLANs is considered. In [11], admission control and resource allocation schemes are proposed for an interworking system which consists of a code division multiple access (CDMA) based cellular network and an IEEE 802.11 distributed coordination function (DCF) based WLAN. In these schemes, total network welfare is maximized to ensure QoS satisfaction in the system. In [34], a resource allocation scheme for an interworking system consisting of an OFDMA based femtocell network and an IEEE 802.11 DCF based WLAN is proposed. Resources of both networks are allocated on the same time-scale, and the WLAN uses basic access scheme with two-way handshaking.

The existing resource allocation schemes have three limitations: (1) they allocate resources of different networks in the interworking system at the same time-scale; (2) they do not fully utilize the QoS support in WLANs; and (3) they do not jointly allocate transmit power levels for different network interfaces at multi-homing capable UEs. Allocating resources of different networks at the same time-scale is not practical as different networks use different resource allocation intervals. To facilitate QoS in WLANs, recent WLAN standards provide contention-based and contention-free polling based channel access mechanisms. These QoS features of WLANs should be considered to maximize the efficiency of the interworking system. Furthermore, jointly allocating transmit power levels for different network interfaces at multi-homing capable UEs is essential for an efficient resource utilization. Joint transmit power allocation is studied in [34] without considering user QoS requirements.

In this chapter, uplink resource allocation for cellular/WLAN interworking to satisfy QoS requirements of multi-homing UEs is investigated. Resources of the two networks in the interworking system are allocated at two different time-scales based on PHY and MAC technologies of the networks. Both contention-based and contention-free channel access schemes of the WLAN are considered, and transmit power of multi-homing UEs is jointly allocated with other network resources.

## 3.3   Cellular/WLAN Interworking System Model

The system model focuses on the first and the second type areas described in Sect. 2.1. Such system is shown in Fig. 3.1, and it consists of a single cell of a cellular network and a WLAN within the coverage of the cell. Uplink resources of this system are allocated. In the system, there are $N$ users belonging to two sets: high-mobility users and low-mobility users. The set of all the users is denoted by $S_N$. The set of low-mobility users within the WLAN coverage is denoted by $S_M$, while the set of remaining users is denoted by $S_S$. For example, $UE_1$ to $UE_4$ in Fig. 3.1 are in $S_M$, while $UE_5$ and $UE_6$ are in $S_S$. Each user has voice and data traffic requirements. All UEs are equipped with WLAN and cellular network interfaces, and have the multi-homing capability. Users in $S_M$ are allowed to simultaneously communicate over both the networks, while users in $S_S$ are only allowed to communicate over the cellular network.

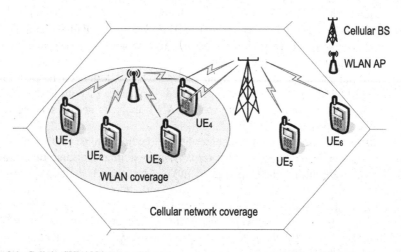

**Fig. 3.1** Cellular/WLAN interworking

The set of subcarriers available at the cellular network is denoted by $\mathcal{K}^C$. Each subcarrier is allocated to only one user in order to avoid CCI among the users. Voice and data traffic services are served through the cellular network. The set of contention-free TXOPs available during a CFP is denoted by $\mathcal{K}^{CF}$. Each contention-free TXOP is allocated to only one user at any given time in order to avoid CCI among the users. Contention-free channel access is more suitable for constant bit rate voice traffic, while contention-based channel access is more suitable for variable bit rate data traffic [32]. To optimize resource utilization subject to QoS requirements in this system, voice traffic is served by contention-free channel access, and data traffic is served by both channel access mechanisms. The sets of users communicate using contention-based and contention-free channel access are denoted by $\mathcal{S}^{CB}$ and $\mathcal{S}^{CF}$ respectively, where $\mathcal{S}^{CB}, \mathcal{S}^{CF} \subseteq \mathcal{S}_M$ and possibly $\mathcal{S}^{CB} \bigcap \mathcal{S}^{CF} \neq \emptyset$.

### 3.3.1 Two-Time-Scale Resource Allocation Framework

Resource allocation intervals of existing cellular networks are shorter than those of the existing WLANs, as cellular networks and WLANs are designed to support high mobility and low mobility users, respectively [12, 13]. Therefore, as shown in Fig. 3.2, resources in the cellular network are allocated at a faster time-scale than the time-scale at which the WLAN resources are allocated. The duration of a time slot in a time-scale is the resource allocation interval of the corresponding network, denoted by $T^L$ and $T^U$ in the fast and slow time-scales ($T^L < T^U$) for the cellular network and the WLAN, respectively. For simplicity, $V_L(= T^U/T^L)$ is assumed to be an integer, and the boundaries of the first time slots in the two time-scales are assumed to be aligned. The resource allocation processes at fast and slow time-scales are referred to as lower and upper levels of the resource allocation process, respectively. Furthermore, as the WLAN resource allocation interval is relatively

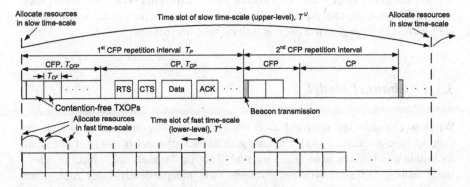

**Fig. 3.2** Resource allocation at slow and fast time-scales

long, to satisfy strict delay and jitter requirements of periodically arriving constant bit rate voice traffic, several short CFPs are utilized within a resource allocation interval of the WLAN instead of using a long CFP [44].

### 3.3.2  Symbols and Notations for the Chapter

The following notations are used for clarity of the symbols in this chapter. The $l$th ($l \in \{0, 1, \ldots, V_L - 1\}$) fast time-scale time slot within the $u$th slow time-scale time slot is referred to as $(u, l)$th time slot. Most of the symbols are written in the form of $X_{i,y}^n$ or $X_{i,y}^n(\cdot)$, where superscript $n$, $n \in \{C, W, CB, CF, L, U\}$, represents the network or the level of resource allocation process. Superscripts $C, W, CB$ and $CF$ denote the cellular network, WLAN, contention-based channel access and contention-free channel access respectively. Superscripts $L$ and $U$ denote lower and upper levels of the resource allocation process respectively. Subscripts denote the user, a particular resource of network $n$, and a time slot. When $n \in \{W, CB\}$, only one subscript which represents the user is used. Boldface letters represent vectors and matrices. Vector $\mathbf{X}$ is defined as $\mathbf{X} = \{X_1, \ldots, X_{|\mathbf{X}|}\}$ with $|\mathbf{X}|$ being the number of elements in $\mathbf{X}$. The optimum of $\mathbf{X}$ is denoted by $\mathbf{X}^*$. The optimum value of variable $X$ is denoted by $X^*$. A decision policy is denoted by $\mathbb{X}$, and the active (i.e., the determined) policy is denoted by $\mathbb{X}^*$.

### 3.3.3  Traffic Model

Traffic generated by each user belongs to two classes: constant bit rate voice traffic and delay tolerant data traffic. Every user always has data traffic to be transmitted. The minimum data rates of voice and data traffic classes required by the $i$th user are denoted by $R_{Vmin,i}$ and $R_{Dmin,i}$, respectively. As voice traffic are highly susceptible to delay and jitter, voice traffic requirements are satisfied in average sense over each slow time-scale time slot. As data traffic are delay tolerant, data traffic requirements are satisfied in average sense over an infinite time horizon.

### 3.3.4  Channel Model

Wireless channels are modeled as a finite-state Markov process to capture the channel time-correlation [45]. Channel gain of each channel is assumed to remain fix within a coherence time ($T_{coh}$) interval. That is, channels are subject to quasi-static fading. Different wireless channels fade independently from each other. Domain of channel gains is partitioned into $K_S$ non-overlapping states. Transition probabilities between different states of a Rayleigh fading channel can be calculated as in [45], assuming that $T^U$ and $T^L$ are shorter than the corresponding channel coherence times to ensure the states do not change within a time slot.

### 3.3.5 Subcarrier and Contention-Free TXOP Allocations, and User Throughputs

Constraints to ensure subcarrier and contention-free TXOP allocations do not cause CCI are derived as follows. Define a variable $\rho_{i,y}^n$ such that $\rho_{i,y}^n = 1$ if the $i$th user is allocated the $y$th resource of network $n \in \{C, CF\}$, i.e., the $y$th OFDM subcarrier or TXOP; and $\rho_{i,y}^n = 0$ otherwise. Then, the constraints to avoid CCI by allocating each resource to only one user is

$$\sum_{i \in S_N} \rho_{i,y}^n \leq 1 \, , \, \forall y \in \mathcal{K}^n. \tag{3.1}$$

Furthermore, $\rho_{i,y}^{CF} = 0, \forall y$ if $i \notin S_M$, as only users in $S_M$ can communicate via the WLAN.

From (2.1) and (2.2), the $i$th user's maximum achievable error free data rate using the $y$th resource of network $n$, $n \in \{C, CF\}$, is given by

$$R_{i,y}^n(P_{i,y}^n) = \sum_{y \in \mathcal{K}^n} \rho_{i,y}^n B \log_2(1 + \alpha_{i,y}^n P_{i,y}^n), \tag{3.2}$$

where $B$ represents the bandwidth of WLAN ($B^W$) or bandwidth of an OFDM subcarrier ($\Delta f = B^C/|\mathcal{K}^C|$); and $B^C$ is the system bandwidth of the cell. Furthermore, as explained in Sect. 2.3.1, $\alpha_{i,y}^{CF} = \alpha_i^W$, $\forall y$ over each channel coherence time interval in the WLAN. Calculation of user throughputs via contention-based channel access of the WLAN is discussed in Sect. 2.3.2.

### 3.3.6 Power Usage of Multi-Homing UEs

The operating time of a UE is governed by the average power (or energy) consumption of uplink communications through WLAN and cellular network interfaces of the UE [10, 46]. Thus, each UE's total average power consumption over each slow time-scale time slot is limited to a predefined maximum. The average power consumption through the WLAN interface for contention-based channel access ($P_{avg,i}^{CB}$ ($\mathbf{P}^{CB}$)) is given by (2.7). Then, total average power consumption of each UE over the $u$th time slot is constrained as follows.

$$P_{avg,i}^{C} + P_{avg,i}^{CB}(\mathbf{P}^{CB}) + \frac{T_{CF}}{T_P} \sum_{j \in \mathcal{K}^{CF}} \rho_{i,j}^{CF} P_{i,j}^{CF} \leq P_{T,i}, \forall i \in \mathcal{S}_N, \tag{3.3}$$

where $P_{avg,i}^{C}$ is the average power usage through the cellular interface during the time slot and $P_{T,i}$ is the total average power available for the $i$th user.

## 3.4  MMDP-Based Optimal Resource Allocation

The objective of resource allocation is to maximize the total throughput of the interworking system subject to satisfaction of QoS requirements. As discussed in Sect. 3.3.1, the resource allocation process consists of upper and lower levels operating at slow and fast time-scales respectively, based on the channel state information. Resources of the WLAN and the cellular network are allocated at the beginnings of the $u$th and the $(u, l)$th time slots respectively, where $u = \{0, 1, 2, \ldots\}$ and $l = \{0, \ldots, V_L - 1\}$. Set of channel gains of the channels between users in $\mathcal{S}_M$ and the WLAN AP at the beginning of the $u$th time slot is referred to as the upper-level state during the $u$th time slot ($\psi_u^U$). Set of channel gains of the channels between all the users and the cellular BS at the beginning of the $(u, l)$th time slot is referred to as the lower-level state during the $(u, l)$th time slot ($\psi_{u,l}^L$). The system state $\{\psi_u^U, \psi_{u,l}^L\}$ is denoted by $\psi_{u,l}$. Sets of all the possible states of upper and lower levels are denoted by $\Psi^U$ and $\Psi^L$, respectively.

The optimal resource allocation process for the cellular/WLAN interworking system is an MMDP [17] due to three reasons: (1) resource allocation process operates at two time-scales, (2) state transition of each level is a Markov process due to Markov channel model, and (3) resource allocations at multiple time slots are jointly optimized to satisfy user QoS requirements as explained in Sect. 3.3.3. The MMDP consists of upper and lower level resource allocation policies [17]. As shown in Fig. 3.3, decisions of the upper-level are made considering the throughputs achieved through and the power consumed at the lower-level. Thus, the upper-level policy ($\mathbb{D}^U$) maps system state $\psi_{u,0}$ to a set of resource allocation decisions ($A_u^U$) at the beginning of $u$th time slot, $u = \{0, 1, 2, \ldots\}$. The lower-level policy ($\mathbb{D}^L$) maps state $\psi_{u,l}^L$ to a set of resource allocation decisions ($A_{u,l}^L$) at the beginning of $(u, l)$th time slot, $l = \{0, \ldots, V_L - 1\}$. Decisions in $A_u^U$ and $A_{u,l}^L$ are $\{P_i^{CB}, P_{i,j}^{CF}, \rho_{i,j}^{CF} | \forall i \in \mathcal{S}_M, j \in \mathcal{K}^{CF}\}$ and $\{P_{i,k}^C, \rho_{i,k}^C | \forall i \in \mathcal{S}_N, k \in \mathcal{K}^C\}$, respectively. The system policy $\{\mathbb{D}^U, \mathbb{D}^L\}$ is denoted by $\mathbb{D}$.

Summation of discounted throughputs (SDTs) [47, 48] over an infinite time horizon is used as the reward or the objective function for the resource allocation. The SDT based reward function reduces susceptibility of the determined decision policies to unpredictable channel changes in the future by giving less importance to those decisions made (and rewards achieved) at far future. The SDTs achieved by the $i$th user at the upper-level over an infinite time horizon with the initial state of

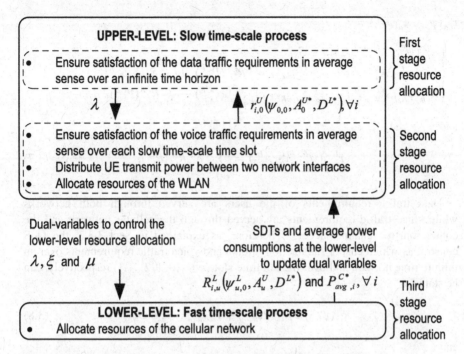

**Fig. 3.3** MMDP-based two-time-scale resource allocation framework

$\psi_{0,0}$ and at the lower-level over the $u$th time slot with the initial state of $\psi_{u,0}^L$ are denoted by $R_i^U(\psi_{0,0}, \mathbb{D})$ and $R_{i,u}^L(\psi_{u,0}^L, A_u^U, \mathbb{D}^L)$, respectively [17]. As the decision policies are stationary (to be discussed), $R_i^U(\psi_{0,0}, \mathbb{D})$ and $R_{i,u}^L(\psi_{u,0}^L, A_u^U, \mathbb{D}^L)$ can be interpreted as the average throughputs that are achieved by the $i$th user over the same periods of time at the upper and lower levels, respectively [47]. They are given by [47, 48]

$$R_i^U(\psi_{0,0}, \mathbb{D}) = \lim_{V_U \to \infty} (1 - \theta) \sum_{u=0}^{V_U - 1} \theta^u r_{i,u}^U(\psi_{u,0}, A_u^U, \mathbb{D}^L) \tag{3.4}$$

and

$$R_{i,u}^L(\psi_{u,0}^L, A_u^U, \mathbb{D}^L) = (1 - \beta) \sum_{l=0}^{V_L - 1} \beta^l r_{i,u,l}^L(\psi_{u,l}^L, A_u^U, A_{u,l}^L), \tag{3.5}$$

where $\theta, \beta \in (0, 1)$ are discount factors. $r_{i,u}^U(\psi_{u,0}, A_u^U, \mathbb{D}^L)$ and $r_{i,u,l}^L(\psi_{u,l}^L, A_u^U, A_{u,l}^L)$ denote throughputs achieved by the $i$th user at the upper and the lower levels during the $u$th and the $(u, l)$th time slots respectively, and are given by [49]

$$r_{i,u}^U(\psi_{u,0}, A_u^U, \mathbb{D}^L) =$$

$$\begin{cases} R_{i,u}^L(\psi_{u,0}^L, A_u^U, \mathbb{D}^L), & \text{if } i \in \mathcal{S}_S; \\ R_{i,u}^L(\psi_{u,0}^L, A_u^U, \mathbb{D}^L) + \frac{T_{CF}}{T_P} \sum_{j \in \mathcal{K}^{CF}} \rho_{i,j}^{CF} R_{i,j}^{CF}(P_{i,j}^{CF}), & \text{if } i \in \mathcal{S}_M \setminus \mathcal{S}^{CB}; \\ R_{i,u}^L(\psi_{u,0}^L, A_u^U, \mathbb{D}^L) + \frac{T_{CF}}{T_P} \sum_{j \in \mathcal{K}^{CF}} \rho_{i,j}^{CF} R_{i,j}^{CF}(P_{i,j}^{CF}) + \frac{T_{CP}}{T_P} R_i^{CB}(\mathbf{P}^{CB}), & \text{if } i \in \mathcal{S}^{CB}; \end{cases}$$

$$(3.6)$$

and

$$r_{i,u,l}^L(\psi_{u,l}^L, A_u^U, A_{u,l}^L) = \sum_{k \in \mathcal{K}^C} \rho_{i,k}^C R_{i,k}^C(P_{i,k}^C). \tag{3.7}$$

Data traffic requirements of the users are served through both networks while voice traffic requirements are served through the cellular network and the contention-free channel access. Therefore, as explained in Sect. 3.3.3, the QoS constraints which ensure satisfaction of data and voice traffic requirements over an infinite time horizon and over the $u$th time slot ($u = \{0, 1, 2, \ldots\}$) respectively, can be stated as

$$R_i^U(\psi_{0,0}, \mathbb{D}) \geq R_{Vmin,i} + R_{Dmin,i}, \ \forall i \in \mathcal{S}_N \tag{3.8}$$

and

$$R_{i,u}^L(\psi_{u,0}^L, A_u^U, \mathbb{D}^L) + \frac{T_{CF}}{T_P} \sum_{j \in \mathcal{K}^{CF}} \rho_{i,j}^{CF} R_{i,j}^{CF}(P_{i,j}^{CF}) \geq R_{Vmin,i}, \ \forall i \in \mathcal{S}_N. \tag{3.9}$$

Since the sum of discounted costs provides a good approximation for the average cost when the policies are stationary [47], $P_{avg,i}^C$ over the $u$th time slot can be calculated via

$$P_{avg,i}^C = (1 - \beta) \sum_{l=0}^{V_L - 1} \beta^l P_{tot,i,l}^C(\psi_{u,l}^L, A_u^U, A_{u,l}^L), \tag{3.10}$$

where $P_{tot,i,l}^C(\psi_{u,l}^L, A_u^U, A_{u,l}^L)$ is the total power allocated by the $i$th user to communicate over the cellular network during the $(u, l)$th time slot, and is also equivalent to $\sum_{k \in \mathcal{K}^C} \rho_{i,k}^C P_{i,k}^C$ over the $(u, l)$th time slot.

The MMDP based optimal resource allocation problem can then be formulated as [17, 47]

$$\mathcal{P}1: \ \max_{\mathbb{D}^U} \max_{\mathbb{D}^L} \ \sum_{i \in \mathcal{S}_N} R_i^U(\psi_{0,0}, \mathbb{D})$$

$$\text{s.t.} \qquad (3.1) \text{ for } n \in \{C, CF\}, (3.3), (3.8) \text{ and } (3.9).$$

To find the optimal $\mathbb{D}^U$ and $\mathbb{D}^L$ solving problem $\mathcal{P}1$, resource allocation should be optimized over three different time intervals: (1) optimize resource allocation over an infinite time horizon subject to (3.8), (2) optimize resource allocation over each upper-level time slot subject to (3.3) and (3.9), and (3) optimize resource allocation over each lower-level time slot. Therefore, $\mathcal{P}1$ is solved in three stages, where the first, the second and the third stages allocate resources over an infinite time horizon, for each upper-level time slot, and for each lower-level time slot, respectively. Resource allocation problem for the $m$th stage ($m = \{2, 3\}$) is derived by decomposing the $(m - 1)$th stage problem into a set of problems, each of which allocates resources over the resource allocation interval of the $m$th stage subject to constraints that must be satisfied within the resource allocation interval of the $m$th stage.

The optimality of the solution, which is obtained via the three stage approach, for $\mathcal{P}1$ is ensured by iterating the $m$th stage ($m = \{1, 2\}$) solution until it reaches the optimal while calculating the optimum $(m + 1)$th stage solution for each $m$th stage iteration. During the iteration process, dual variables of the $m$th stage are passed to the $(m + 1)$th stage while throughputs/SDTs achieved and power consumed at the $(m + 1)$th stage are feedback to the $m$th stage, as shown in Fig. 3.3. At the $(m + 1)$th stage, the received dual variables are utilized to configure objective function of this stage resource allocation problem such that the $(m + 1)$th stage assists maximizing the $m$th stage objective. At the $m$th stage, the received information is utilized to update the dual variables.

Problem $\mathcal{P}1$ is a non-convex optimization problem. Therefore, to reduce the computational complexity involved with solving $\mathcal{P}1$, $\mathcal{P}1$ is relaxed. Due to the relaxations, the policies determined in this chapter ($\mathbb{D}^{U*}$ and $\mathbb{D}^{L*}$) are not optimal for $\mathcal{P}1$ in certain scenarios. Therefore, $\mathbb{D}^{U*}$ and $\mathbb{D}^{L*}$ are referred to as active or determined upper and lower level decision policies, respectively. Derivations of $\mathbb{D}^{U*}$, which is found by solving the first and the second stage resource allocation problems, and $\mathbb{D}^{L*}$, which is found by solving the third stage resource allocation problem, are discussed in Sects. 3.5 and 3.6, respectively.

Using the state transition probabilities calculated based on channel statistics, $\mathbb{D}^{U*}$ and $\mathbb{D}^{L*}$ can be determined in advance and applied to the interworking system based on the initial states. The applied policies select $A_u^{U*}$ and $A_{u,l}^{L*}$ for the $u$th and the $(u, l)$th time slots respectively, based on the states of the two levels during the time slots. The policies $\mathbb{D}^{U*}$ and $\mathbb{D}^{L*}$ are required to be recalculated when the channel statistics or the number of users in the system or user QoS requirements change.

## 3.5 Upper-Level of MMDP-Based Optimal Resource Allocation

Upper-level resource allocation policy $\mathbb{D}^{U*}$ is determined as follows. Total SDT at the upper-level is maximized subject to satisfaction of (3.8) over an infinite time horizon with initial system state of $\psi_{0,0}$; this first stage problem is denoted by

$\mathcal{P}2$. $\mathcal{P}2$ is a convex optimization problem and it can be solved by solving the dual problem [50]. To find the dual function which is the minimum of the Lagrangian, first the Lagrangian is decomposed into a set of terms, each of which is a *negative* summation of weighted user throughputs corresponding to one time slot. Second, to find the minimum of the Lagrangian, $A_u^{U*}$ for the $u$th time slot ($u = \{0, 1, 2, \ldots\}$) is determined such that it maximizes the summation of weighted user throughputs corresponding to the $u$th time slot subject to satisfaction of (3.1) for $n = CF$, (3.3) and (3.9). The problem which finds $A_u^{U*}$ for the $u$th time slot is referred to as the second stage problem, and is denoted by $\mathcal{P}3$. First and second stage problems are solved in Sects. 3.5.1 and 3.5.2, respectively. Furthermore, the conditions which the third stage resource allocation at the lower-level should satisfy to ensure optimality of the three stage solution for $\mathcal{P}1$ are derived in Sect. 3.5.2.

### 3.5.1   Optimization of Resource Allocations over an Infinite Time Horizon

First stage resource allocation problem can be stated as

$$\mathcal{P}2 : \max_{\mathbb{D}^U} \quad \sum_{i \in \mathcal{S}_N} R_i^U(\psi_{0,0}, \mathbb{D}^U, \mathbb{D}^{L*})$$

$$\text{s.t.} \quad C1 : (3.8).$$

The active policy $\mathbb{D}^{L*}$ is used in $\mathcal{P}2$, as $\mathbb{D}^{L*}$ is calculated by solving the third stage resource allocation problem for each iteration of the algorithm which solves $\mathcal{P}2$. From (3.4) to (3.7), the objective function of $\mathcal{P}2$ is a concave function, and the feasible region is a convex set. Therefore, $\mathcal{P}2$ is a convex optimization problem, and is solved by maximizing the dual function which is obtained by minimizing the Lagrangian of $\mathcal{P}2$ with respect to $\mathbb{D}^U$ [50]. The Lagrangian of $\mathcal{P}2$ is

$$L^U(\psi_{0,0}, \lambda, \mathbb{D}^U, \mathbb{D}^{L*}) = \sum_{i \in \mathcal{S}_N} \left[ \lambda_i(R_{Vmin,i} + R_{Dmin,i}) - (1 + \lambda_i) R_i^U(\psi_{0,0}, \mathbb{D}^U, \mathbb{D}^{L*}) \right],$$

$$(3.11)$$

where $\lambda_i$, $\forall i$ are dual variables.

The iterative algorithm which solves $\mathcal{P}2$ can be summarized as follows. First, $\lambda$ is initialized (e.g., $\lambda \leftarrow \{0, \ldots, 0\}$). Second, $\mathbb{D}^U$ which minimizes $L^U(\psi_{0,0}, \lambda, \mathbb{D}^U, \mathbb{D}^{L*})$ for $\lambda$ is found. To update $\lambda$ for the next iteration, $R_i^U(\psi_{0,0}, \mathbb{D}^U, \mathbb{D}^{L*})$, $\forall i$ are also found in this step. Third, $\lambda$ is adjusted toward $\lambda^*$ using the subgradient method [10, 51, 52]. The second and the third steps are repeated until $\lambda$ reaches $\lambda^*$. When $\lambda$ reaches $\lambda^*$, each $\lambda_i$ satisfies the complementary slackness condition [50] and $\mathbb{D}^{U*}$ has been found.

To implement the algorithm which solves $\mathcal{P}2$, $\mathbb{D}^U$ and $R_i^U(\psi_{0,0}, \mathbb{D}^U, \mathbb{D}^{L*})$, $\forall i$ for any $\lambda$ can be calculated as follows. From (3.11) and since $\sum_{i \in \mathcal{S}_N} \lambda_i(R_{Vmin,i} + R_{Dmin,i})$ does not depend on $\mathbb{D}^U$, $\mathbb{D}^U$ is determined such that it maximizes $\sum_{i \in \mathcal{S}_N}(1 + \lambda_i)R_i^U(\psi_{0,0}, \mathbb{D}^U, \mathbb{D}^{L*})$. When $\sum_{i \in \mathcal{S}_N}(1 + \lambda_i)R_i^U(\psi_{0,0}, \mathbb{D}^U, \mathbb{D}^{L*})$ is maximized, by (3.4) and using the Bellman optimality equation [47], it is given by the optimality equation shown below.

$$L_{sup}^U(\psi_{0,0}, \boldsymbol{\lambda}) = (1 - \theta) \max_{A_0^U} \left[ \sum_{i \in \mathcal{S}_N}(1 + \lambda_i) r_{i,0}^U(\psi_{0,0}, A_0^U, \mathbb{D}^{L*}) \right]$$
$$+ \theta \sum_{\psi_1^U \in \Psi^U} \sum_{\psi_{1,0}^L \in \Psi^L} P_{\psi_0^U \psi_1^U}^U P_{\psi_{0,0}^L \psi_{1,0}^L}^L L_{sup}^U(\psi_{1,0}, \boldsymbol{\lambda}) \tag{3.12}$$

with

$$L_{sup}^U(\psi_{u,0}, \boldsymbol{\lambda}) = \sup_{\mathbb{D}^U} \left[ \sum_{i \in \mathcal{S}_N}(1 + \lambda_i) R_i^U(\psi_{u,0}, \mathbb{D}^U, \mathbb{D}^{L*}) \right],$$

where $P_{\psi_0^U \psi_1^U}^U$ and $P_{\psi_{0,0}^L \psi_{1,0}^L}^L$ are probabilities of the upper and the lower level states change from $\psi_0^U$ to $\psi_1^U$ and from $\psi_{0,0}^L$ to $\psi_{1,0}^L$ at the end of the 0th time slot, respectively. Equation (3.12) is a recursive equation, and $A_u^{U*}$ for the $u$th time slot is determined such that the summation of weighted throughputs given by $\sum_{i \in \mathcal{S}_N}(1 + \lambda_i) r_{i,u}^U(\psi_{u,0}, A_u^U, \mathbb{D}^{L*})$ is maximized [48]. Moreover, the resource allocation problems corresponding to different time slots are independent of each other. As $\mathbb{D}^{U*}$ is a stationary policy (to be discussed in Sect. 3.5.2), finding $A_0^{U*}$ for each $\psi_{0,0} \in \{\Psi^U, \Psi^L\}$ at the 0th time slot is sufficient to find $\mathbb{D}^{U*}$. Then, $R_i^U(\psi_{0,0}, \mathbb{D}^U, \mathbb{D}^{L*})$ can be determined by solving the Bellman optimality equation for the $i$th user written using (3.4). Bellman optimality equation solving methods, such as Value Iteration algorithm and its variants, are explained in [48].

### 3.5.2 Resource Allocation for an Upper-Level Time Slot

Second stage resource allocation problem is solved to find $A_u^{U*}$ at system state $\psi_{u,0}$ during the $u$th time slot, such that $A_u^{U*}$ maximizes $\sum_{i \in \mathcal{S}_N}(1 + \lambda_i) r_{i,u}^U(\psi_{u,0}, A_u^U, \mathbb{D}^{L*})$ subject to (3.1) for $n = CF$, (3.3) and (3.9). This problem is a non-convex optimization problem due to the integer constraint imposed on $\rho_{i,j}^{CF}$ (see Sect. 3.3.5). Therefore, to reduce the computational complexity involved with solving the problem, the problem is relaxed to be a convex optimization problem by relaxing the integer constraint such that $\rho_{i,j}^{CF} \in [0,1]$, [53]. To calculate power usage and throughputs over partially allocated TXOPs, following two substitutions are defined.

$$\bar{P}_{i,j}^{CF} = \rho_{i,j}^{CF} P_{i,j}^{CF} \tag{3.13}$$

and

$$\bar{R}_{i,j}^{CF}(\bar{P}_{i,j}^{CF}, \rho_{i,j}^{CF}) = \left(\frac{T_{CF}}{T_P}\right)\rho_{i,j}^{CF} R_{i,j}^{CF}(\bar{P}_{i,j}^{CF}/\rho_{i,j}^{CF}), \tag{3.14}$$

where $\bar{R}_{i,j}^{CF}(\bar{P}_{i,j}^{CF}, \rho_{i,j}^{CF})$ is a concave function [49]. For notation simplicity, the substitution below is defined.

$$\bar{R}_i^{CB}(\mathbf{P}^{CB}) = \left(\frac{T_{CP}}{T_P}\right)R_i^{CB}(\mathbf{P}^{CB}). \tag{3.15}$$

With substitutions from (3.6) and (3.13)–(3.15) to the second stage resource allocation problem, the relaxed problem can be stated as

$$\mathcal{P}3: \max_{A_u^U} \quad \sum_{i \in \mathcal{S}_N}(1+\lambda_i)\left[R_{i,u}^L(\psi_{u,0}^L, A_u^U, \mathbb{D}^{L*}) + \sum_{j \in \mathcal{K}^{CF}} \bar{R}_{i,j}^{CF}(\bar{P}_{i,j}^{CF}, \rho_{i,j}^{CF})\right]$$

$$+ \sum_{i \in \mathcal{S}^{CB*}}(1+\lambda_i)\bar{R}_i^{CB}(\mathbf{P}^{CB})$$

s.t.    C2 : $\sum_{i \in \mathcal{S}_M}\rho_{i,j}^{CF} \leq 1$, $\forall j \in \mathcal{K}^{CF}$

C3 : $R_{i,u}^L(\psi_{u,0}^L, A_u^U, \mathbb{D}^{L*}) + \sum_{j \in \mathcal{K}^{CF}} \bar{R}_{i,j}^{CF}(\bar{P}_{i,j}^{CF}, \rho_{i,j}^{CF}) \geq R_{Vmin,i}$, $\forall i \in \mathcal{S}_N$

C4 : $P_{avg,i}^{CB}(\mathbf{P}^{CB}) + P_{avg,i}^C + \frac{T_{CF}}{T_P}\sum_{j \in \mathcal{K}^{CF}} \bar{P}_{i,j}^{CF} \leq P_{T,i}$, $\forall i \in \mathcal{S}_N$

C5 : $0 \leq \rho_{i,j}^{CF} \leq 1$, $\forall i \in \mathcal{S}_M, j \in \mathcal{K}^{CF}$

C6 : $\bar{P}_{i,j}^{CF} \geq 0$, $P_i^{CB} \geq 0$, $\forall i \in \mathcal{S}_N, j \in \mathcal{K}^{CF}$.

Problem $\mathcal{P}3$ is a convex optimization problem. Convexity of C4, i.e., convexity of the set $\{P_{avg,i}^C, \bar{P}_{i,j}^{CF}, P_i^{CB}|\text{C4 is satisfied}, i \in \mathcal{S}_N, j \in \mathcal{K}^{CF}\}$, is proved in Appendix A.2.

Next, the relationship between $\mathcal{P}3$ and the third stage resource allocation problem which determines $\mathbb{D}^{L*}$ for the lower-level is illustrated. Then, $A_u^{U*}$ is derived by solving $\mathcal{P}3$ using Karush-Kuhn-Tucker (KKT) conditions [50]. The Lagrangian for $\mathcal{P}3$ can be written as

$$L^{U(2)}(A_u^U, \gamma, \xi, \mu) = -\sum_{i \in \mathcal{S}_N} (1 + \lambda_i + \xi_i) \Big[ R_{i,u}^L(\psi_{u,0}^L, A_u^U, \mathbb{D}^{L*}) + \sum_{j \in \mathcal{K}^{CF}} \bar{R}_{i,j}^{CF}(\bar{P}_{i,j}^{CF}, \rho_{i,j}^{CF}) \Big]$$

$$+ \sum_{i \in \mathcal{S}_N} \Big[ \sum_{j \in \mathcal{K}^{CF}} \gamma_j \rho_{i,j}^{CF} + \xi_i R_{Vmin,i} + \mu_i \Big( P_{avg,i}^{CB}(\mathbf{P}^{CB}) + P_{avg,i}^C + \frac{T_{CF}}{T_P} \sum_{j \in \mathcal{K}^{CF}} \bar{P}_{i,j}^{CF} - P_{T,i} \Big) \Big]$$

$$- \sum_{i \in \mathcal{S}^{CB*}} (1 + \lambda_i) \bar{R}_i^{CB}(\mathbf{P}^{CB}) - \sum_{j \in \mathcal{K}^{CF}} \gamma_j, \qquad (3.16)$$

where $\gamma_j$, $\xi_i$ and $\mu_i$, $\forall i,j$ are the dual variables associated with C2, C3 and C4, respectively. As the optimal solution for problem $\mathcal{P}3$ minimizes (3.16) subject to C5 and C6, $\mathbb{D}^{L*}$ is determined such that it maximizes $\sum_{i \in \mathcal{S}_N} (1 + \lambda_i + \xi_i) R_{i,u}^L(\psi_{u,0}^L, A_u^U, \mathbb{D}^L)$. Furthermore, $\mathbb{D}^{L*}$ should satisfy the following KKT condition of $\mathcal{P}3$ to ensure the optimality of the solution, which is obtained using the three stages, for $\mathcal{P}1$.

$$(1 + \lambda_i + \xi_i) \frac{\partial R_{i,u}^L(\psi_{u,0}^L, A_u^U, \mathbb{D}^{L*})}{\partial P_{avg,i}^C}\Big|_{P_{avg,i}^C = P_{avg,i}^{C*}} \begin{cases} = \mu_i, & \text{if } P_{avg,i}^{C*} > 0 ; \\ < \mu_i, & \text{otherwise}; \end{cases} \quad \forall i \in \mathcal{S}_N.$$

$$(3.17)$$

Dual variables $\mu$ and $\xi$ couple the upper and the lower level resource allocations to optimally distribute the transmit power available at the UEs among WLAN and cellular network interfaces and to optimally utilize the resources of the two networks to satisfy the users' voice traffic requirements, respectively. Due to this coupling, once resources of the lower-level are allocated, achieved SDTs (i.e., $R_{i,u}^L(\psi_{u,0}^L, A_u^U, \mathbb{D}^{L*})$, $\forall i$) and the average power consumptions at the lower-level (i.e., $P_{avg,i}^{C*}$, $\forall i$) are feedback to the upper-level to solve $\mathcal{P}3$, as shown in Fig. 3.3.

Problem $\mathcal{P}3$ is solved to find $A_u^{U*}$ as follows. First, $\xi$ and $\mu$ are initialized. Second, the optimal allocations of contention-free TXOPs, UE transmit power levels during contention-free and contention-based TXOPs, $R_{i,u}^L(\psi_{u,0}^L, A_u^U, \mathbb{D}^{L*})$, $\forall i$ and $P_{avg,i}^{C*}$, $\forall i$ are calculated based on $\xi$ and $\mu$. Third, $\mu$ is updated toward $\mu^*$ using the subgradient method [10, 51, 52]. The second and third steps are repeated until $\mu^*$ is found. Forth, $\xi$ is updated toward $\xi^*$ using the subgradient method. The last three steps are repeated until $\xi^*$ is found.

In the remaining of this section, the decision set $A_u^{U*}(= \{P_i^{CB*}, \bar{P}_{i,j}^{CF*}, \rho_{i,j}^{CF*} | \forall i \in \mathcal{S}_M, j \in \mathcal{K}^{CF}\})$ is derived by solving $\mathcal{P}3$, and explained how $\mathcal{S}^{CB*}$ is determined. In addition, the optimality of $A_u^U$ for the initial problem (i.e., the problem prior to the relaxation) is also discussed.

### 3.5.2.1  Allocations of Contention-Free TXOPs and Transmit Power Levels

Based on the KKT conditions for $\mathcal{P}3$, the optimal transmit power levels of the users during contention-free TXOPs are given by

$$\bar{P}_{i,j}^{CF*} = \rho_{i,j}^{CF*}\Theta_i , \ \forall i \in \mathcal{S}_M, j \in \mathcal{K}^{CF}, \tag{3.18}$$

where

$$\Theta_i = \left[\frac{B^W}{\ln(2)}\frac{(1+\lambda_i+\xi_i^*)}{\mu_i^*} - \frac{1}{\alpha_i^W}\right]^+$$

and $[x]^+ = \max\{0, x\}$. Then, the optimal contention-free TXOP allocation can be calculated as follows. Let

$$\Gamma_{i,j} = (1+\lambda_i+\xi_i^*)\cdot\frac{\partial\bar{R}_{i,j}^{CF}(\bar{P}_{i,j}^{CF}, \rho_{i,j}^{CF})}{\partial\rho_{i,j}^{CF}}\Big|_{\bar{P}_{i,j}^{CF}=\bar{P}_{i,j}^{CF*}}$$

$$= \frac{(1+\lambda_i+\xi_i^*)T_{CF}B^W}{T_P}\left[\log_2(1+\alpha_i^W\Theta_i) - \frac{1}{\ln(2)}\frac{\alpha_i^W\Theta_i}{1+\alpha_i^W\Theta_i}\right], \ \forall i \in \mathcal{S}_M, j \in \mathcal{K}^{CF}. \tag{3.19}$$

Since $\Gamma_{i,j}$ is independent of $\rho_{i,j}^{CF}$ and from the KKT conditions, the $j$th TXOP is allocated to the user with the largest $\Gamma_{i,j}$ [49]. However, when there are multiple users with their $\Gamma_{i,j}$ values equal to the largest $\Gamma_{i,j}$ for the $j$th TXOP, the optimal solution for $\mathcal{P}3$ allocates fractions of the TXOP among these users allowing them to time-share the TXOP.

From (3.19), $\Gamma_{i,j}$ values of the $i$th user for all the TXOPs within the $u$th time slot are the same, as the channel gain (or $\alpha_i^W$) and $\Theta_i$ remain unchanged over the $u$th time slot. As a result, the $i$th user is either allocated the same fraction from each TXOP or allocated all the TXOPs. When there are $N'$ users $\{i_1, i_2, \ldots, i_{N'}\}$ with their $\Gamma_{i,j}$ values equal to the largest, the optimal fractional values for $\rho_{i,j}^{CF}, i = \{i_1, i_2, \ldots, i_{N'}\}$ are determined based on the primal feasibility of those $\rho_{i,j}^{CF}$'s with respect to C2, C3 and C4. That is, the optimal set of $\rho_{i,j}^{CF}, i = \{i_1, i_2, \ldots, i_{N'}\}$ is a solution which satisfies C2 with equality and satisfies the following set of linear inequalities.

$$\rho_{i,j}^{CF}|\mathcal{K}^{CF}|\frac{T_{CF}B^W\log_2(1+\alpha_i^W\Theta_i)}{T_P} \geq R_{Vmin,i} - R_{i,u}^L(\psi_{u,0}^L, A_u^U, \mathbb{D}^{L*}) , \ i = \{i_1, i_2, \ldots, i_{N'}\} \tag{3.20}$$

and

$$\rho_{i,j}^{CF}|\mathcal{K}^{CF}|\Theta_i\frac{T_{CF}}{T_P} \leq P_{T,i} - P_{avg,i}^{CB}(\mathbf{P}^{CB}) - P_{avg,i}^{C} , \ i = \{i_1, i_2, \ldots, i_{N'}\}. \tag{3.21}$$

As the objective is to allocate resources based on the PHY and the MAC technologies of the networks, a near optimal TXOP allocation for the initial problem is found by rounding $\rho_{i,j}^{CF}|\mathcal{K}^{CF}|$ values to the nearest integers. The rounded values indicate the number of TXOPs allocated to each user. Furthermore, if $\rho_{i,j}^{CF}|\mathcal{K}^{CF}|$, $\forall i$ are integers, they are the optimal TXOP allocation for the initial problem.

### 3.5.2.2 Allocations of Users and Transmit Power for Contention-Based Access

Second stage resource allocation problem should be formulated as a convex optimization problem in order to reduce the required computational capacity to solve the problem. However, $R_i^{CB}(\mathbf{P}^{CB})$ given by (2.5) is a non-concave function when $N_W$ varies. Thus, to formulate the second stage problem as a convex optimization problem, $\mathcal{S}^{CB*}$ should be determined prior to allocating the other upper-level resources. In the MMDP based resource allocation algorithm, $\mathcal{S}^{CB*}$ which achieves the highest total SDT at the upper-level is found via searching over $\mathcal{S}_M$. A low complexity method to find a near optimal $\mathcal{S}^{CB}$ is described in Sect. 3.7.

From (2.5), it can be seen that $\bar{R}_i^{CB}(\mathbf{P}^{CB})$ depends not only on the $i$th user's transmit power level, but also on the transmit power levels of the other users in $\mathcal{S}^{CB*}$. Therefore, $P_i^{CB*}$, $\forall i \in \mathcal{S}^{CB*}$ are correlated. Based on the KKT conditions for $\mathcal{P}3$, $P_i^{CB*} > 0$ only if

$$\frac{\partial \bar{R}_i^{CB}(\mathbf{P}^{CB})}{\partial P_i^{CB}}\bigg|_{\substack{\mathbf{P}_{-i}^{CB}=\mathbf{P}_{-i}^{CB*} \\ P_i^{CB}=0}} > \frac{\mu_i^*}{1+\lambda_i} \frac{\partial P_{avg,i}^{CB}(\mathbf{P}^{CB})}{\partial P_i^{CB}}\bigg|_{\substack{\mathbf{P}_{-i}^{CB}=\mathbf{P}_{-i}^{CB*} \\ P_i^{CB}=0}}, \tag{3.22}$$

where $\mathbf{P}_{-i}^{CB}$ is a vector which consists of the power levels of the users in $\mathcal{S}^{CB*}$ except the $i$th user. As these partial derivatives are not defined when $P_i^{CB} = 0$, (3.22) is rewritten by taking the limits of the partial derivatives as $P_i^{CB} \to 0$. Then, (3.22) reduces to

$$P_i^{CB*} \begin{cases} > 0, & \text{if } \frac{B^W \alpha_i^W}{\ln(2)} > \frac{\mu_i^*}{1+\lambda_i}; \\ = 0, & \text{otherwise}; \end{cases} \quad \forall i \in \mathcal{S}^{CB*}. \tag{3.23}$$

Furthermore, for the case of $P_i^{CB*} > 0$, the two sides of (3.22) become equal when the partial derivatives are evaluated at $P_i^{CB} = P_i^{CB*}$. Therefore, when $P_i^{CB*} > 0$, $P_i^{CB*}$ can be found by solving (3.24) shown below, using Newton's method if $\mathbf{P}_{-i}^{CB*}$ is known [54].

$$\frac{1+\lambda_i}{\mu_i^*} \cdot \frac{N_W \alpha_i^W \bar{R}_i^{CB}(\mathbf{P}^{CB*})}{1+P_i^{CB*}\alpha_i^W} = \frac{T_{CP}}{T_P}\left[ \ln(1+P_i^{CB*}\alpha_i^W) - \frac{P_i^{CB*}\alpha_i^W}{1+P_i^{CB*}\alpha_i^W} \right]$$
$$+ \frac{\ln(2)N_W P_i^{CB*}\alpha_i^W \bar{R}_i^{CB}(\mathbf{P}^{CB*})}{B^W \cdot (1+P_i^{CB*}\alpha_i^W)\ln(1+P_i^{CB*}\alpha_i^W)}. \tag{3.24}$$

Existence of a solution for (3.24) is proved in Appendix A.3.

Since $P_i^{CB*}, \forall i \in \mathcal{S}^{CB*}$ are correlated, $\mathbf{P}^{CB*}$ is found using an iterative algorithm. In each iteration, $P_i^{CB*}, \forall i \in \mathcal{S}^{CB*}$ are calculated by (3.24) using the $\mathbf{P}^{CB*}$ calculated in the previous iteration. The algorithm terminates when the changes to $P_i^{CB*}, \forall i \in \mathcal{S}^{CB*}$ are negligible. Convergence of this iterative algorithm is proved in Appendix A.4.

The policy $\mathbb{D}^{U*}$ is a stationary policy. Equations (3.18)–(3.24) show that $A_u^{U*}$ is independent of the time slot. Therefore, $A_u^{U*}$ which is determined for the $u$th time slot can be used at the $v$th time slot ($v = \{0,1,2,\ldots\}$) when states during the $u$th and the $v$th time slots are equivalent (i.e., $\psi_{u,0} = \psi_{v,0}$). Thus, $\mathbb{D}^{U*}$ is a stationary policy [48]. The algorithm to determine $\mathbb{D}^{U*}$ for a given initial state $\psi_{0,0}$ is shown in Algorithm 1.

---

**Algorithm 1** : Upper-Level Policy

---

**input**    : $\{\psi_0^U, \psi_{0,0}^L\}$, $\mathcal{S}_M$, $\mathcal{S}_S$, and $P_{T,i}$, $R_{Vmin,i}$ and $R_{Dmin,i}$, $\forall i$
**output**  : $A_u^{U*} = \{P_i^{CB*}, \rho_{i,j}^{CF*}, \bar{P}_{i,j}^{CF*} | \forall i \in \mathcal{S}_M, j \in \mathcal{K}^{CF}\}$ for every $\{\psi^U, \psi^L\} \in \{\Psi^U, \Psi^L\}$, and $\mathcal{S}^{CB*}$

**while** $\mathcal{S}^{CB}$ *is not optimal* **do**
  $\lambda \leftarrow \{0,\ldots,0\}$;
  **while** $\lambda$ *is not optimal* **do**
    **for** *each* $\{\psi^U, \psi^L\} \in \{\Psi^U, \Psi^L\}$ **do**
      $\xi \leftarrow \{0,\ldots,0\}$ and $\mu \leftarrow \{\mu_1,\ldots,\mu_N\}$;
      **while** $\xi$ *is not optimal* **do**
        **while** $\mu$ *is not optimal* **do**
          Calculate $\bar{P}_{i,j}^{CF*}$, $\rho_{i,j}^{CF*}$ and $P_i^{CB*}$ by (3.18)–(3.21), (3.23) and (3.24) at state $\psi^U$;
          Allocate resources at the Lower-Level by Algorithm 2 when the initial state is $\psi^L$;
          Update $\mu_i$, $\forall i$;
        **end**
        Update $\xi_i$, $\forall i$;
      **end**
    **end**
    For each user, calculate $R_i^U(\psi_{0,0}, \mathbb{D})$ by solving the Bellman optimality equation written using (3.4);
    Update $\lambda_i$, $\forall i$;
  **end**
  Calculate the total SDT at the upper-level, $\sum_{i=1}^N R_i^U(\psi_{0,0}, \mathbb{D})$;
**end**

---

## 3.6  Lower-Level of MMDP-Based Optimal Resource Allocation

The lower-level allocates resources to maximize $\sum_{i \in S_N} (1+\lambda_i+\xi_i) R_{i,u}^L(\psi_{u,0}^L, A_u^U, \mathbb{D}^L)$ over the $u$th time slot subject to (3.1) for $n = C$ and (3.17) (see Sect. 3.5.2). To this end, $\mathbb{D}^{L*}$ is determined as follows. First, the resource allocation problem over the $u$th time slot is decomposed to a set of independent subproblems, each of which allocates resources for one lower-level time slot. Second, $A_{u,l}^{L*}$ for the $(u, l)$th time slot, $l = \{0, 1, \ldots, V_L - 1\}$, is found by solving the subproblem which corresponds to the same time slot; this third stage resource allocation problem is denoted by $\mathcal{P}4$ (see Sect. 3.4). Based on $A_{u,l}^{L*}$ found for the $(u, l)$th time slot, it is shown that $\mathbb{D}^{L*}$ is a stationary policy.

To decompose the resource allocation problem which maximizes $\sum_{i \in S_N} (1 + \lambda_i + \xi_i) R_{i,u}^L(\psi_{u,0}^L, A_u^U, \mathbb{D}^L)$ over the $u$th time slot to a set of independent subproblems, the Bellman optimality equation written for the lower-level using (3.5) with the assumption of $V_L$ is very large is used. This assumption is reasonable as $T^L \ll T^U$. Then, the Bellman optimality equation is given by [12, 13]

$$\sum_{i \in S_N} (1 + \lambda_i + \xi_i) R_{i,u}^L(\psi_{u,0}^L, A_u^U, \mathbb{D}^{L*})$$

$$= (1 - \beta) \max_{A_{u,0}^L} \left[ \sum_{i \in S_N} (1 + \lambda_i + \xi_i) r_{i,u,0}^L(\psi_{u,0}^L, A_u^U, A_{u,0}^L) \right]$$

$$+ \beta \sum_{\psi_{u,1}^L \in \Psi^L} P_{\psi_{u,0}^L \psi_{u,1}^L}^{L(2)} \sum_{i \in S_N} (1 + \lambda_i + \xi_i) R_{i,u}^L(\psi_{u,1}^L, A_u^U, \mathbb{D}^{L*}), \qquad (3.25)$$

where $P_{\psi_{u,0}^L \psi_{u,1}^L}^{L(2)}$ is the probability of lower-level state changes from $\psi_{u,0}^L$ to $\psi_{u,1}^L$ at the end of the $(u, 0)$th time slot. As the left hand side of (3.25) is maximized when $A_{u,l}^{L*}$ for the $(u, l)$th time slot maximizes $\sum_{i \in S_N} (1 + \lambda_i + \xi_i) r_{i,u,l}^L(\psi_{u,l}^L, A_u^U, A_{u,l}^L)$, the lower-level resources are allocated for the $(u, l)$th time slot such that $\sum_{i \in S_N} (1+\lambda_i+\xi_i) r_{i,u,l}^L(\psi_{u,l}^L, A_u^U, A_{u,l}^L)$ is maximized [48]. The lower level resources are subcarriers and user transmit power. Furthermore, these resource allocation subproblems corresponding to the lower-level time slots are independent of each other.

Similar to the non-convexity caused by the integer constraint imposed on $\rho_{i,j}^{CF}$ (see Sect. 3.5.2), the integer constraint imposed on $\rho_{i,k}^C$ makes the subproblem corresponding to the $(u, l)$th time slot non-convex (see Sect. 3.3.5). To reduce the computational capacity required to solve the subproblem, the subproblem is relaxed by following the same relaxation process used in Sect. 3.5.2; that is, $\rho_{i,k}^C \in [0, 1]$, $\bar{P}_{i,k}^C = \rho_{i,k}^C P_{i,k}^C$ and $\bar{R}_{i,k}^C(\bar{P}_{i,k}^C, \rho_{i,k}^C) = \rho_{i,k}^C R_{i,k}^C(\bar{P}_{i,k}^C / \rho_{i,k}^C)$. The relaxed subproblem is considered as problem $\mathcal{P}4$.

Since $\mathcal{P}4$ is solved subject to (3.17) over the $u$th time slot, (3.17) is first translated into a set of constraints, each corresponds to one lower-level time slot, by substituting (3.5), (3.7) and (3.10) into (3.17). Then, the constraint corresponding to the $(u, l)$th time slot is given by

$$(1 + \lambda_i + \xi_i) \frac{\partial \bar{R}_{i,k}^C(\bar{P}_{i,k}^C, \rho_{i,k}^C)}{\partial \bar{P}_{i,k}^C} \bigg|_{\bar{P}_{i,k}^C = \bar{P}_{i,k}^{C*}} \begin{cases} = \mu_i, & \text{if } \bar{P}_{i,k}^{C*} > 0; \\ < \mu_i, & \text{otherwise}; \end{cases} \quad \forall i \in \mathcal{S}_N, k \in \mathcal{K}^C.$$

(3.26)

From (3.26),

$$\bar{P}_{i,k}^{C*} = \rho_{i,k}^{C*} \left[ \frac{\Delta f}{\ln(2)} \frac{(1 + \lambda_i + \xi_i)}{\mu_i} - \frac{1}{\alpha_{i,k}^C} \right]^+.$$

(3.27)

The remaining subcarrier allocation problem for the $(u, l)$th time slot can be stated as follows.

$$\mathcal{P}5 : \max_{\rho^C} \quad \sum_{i \in \mathcal{S}_N} \sum_{k \in \mathcal{K}^C} (1 + \lambda_i + \xi_i) \bar{R}_{i,k}^C(\bar{P}_{i,k}^C, \rho_{i,k}^C)$$

$$\text{s.t.} \quad C7 : \sum_{i \in \mathcal{S}_N} \rho_{i,k}^C \le 1, \ \forall k \in \mathcal{K}^C$$

$$C8 : 0 \le \rho_{i,k}^C \le 1, \ \forall i \in \mathcal{S}_N, k \in \mathcal{K}^C.$$

Problem $\mathcal{P}5$ is a convex optimization problem. Therefore, from (3.27) and the KKT conditions for $\mathcal{P}5$, and using the same approach used for deriving $\rho_{i,j}^{CF*}$ in Sect. 3.5.2.1, the optimal subcarrier allocation is given by [49]

$$\rho_{i',k}^{C*} = \begin{cases} 1, & \text{if } i' = \arg\max_{\forall i} \{\Lambda_{i,k}\}; \\ 0, & \text{otherwise}; \end{cases} \quad \forall i' \in \mathcal{S}_N, k \in \mathcal{K}^C,$$

(3.28)

where

$$\Lambda_{i,k} = (1 + \lambda_i + \xi_i) \left[ \log_2(1 + \alpha_{i,k}^C P_{i,k}^{C*}) - \frac{1}{\ln(2)} \frac{\alpha_{i,k}^C P_{i,k}^{C*}}{1 + \alpha_{i,k}^C P_{i,k}^{C*}} \right].$$

When there are multiple users with their $\Lambda_{i,k}$ values equivalent to the largest $\Lambda_{i,k}$ for the $k$th subcarrier, the optimal solution for $\mathcal{P}5$ requires allocation of fractions, which satisfy C7 with equality, of the $k$th subcarrier among these users. When such equality of $\Lambda_{i,k}$ occurs, the resource allocation algorithm proposed in this chapter randomly allocates the $k$th subcarrier to one of the users with the largest $\Lambda_{i,k}$ as fractional subcarrier allocations are not supported by the PHY. Random subcarrier allocation in this scenario does not significantly deviate the system throughput/QoS performance from the optimum due to two reasons: (1) subcarrier bandwidth is small as there are a large number of subcarriers; (2) probability of multiple users having equivalent $\Lambda_{i,k}$ values for more than one subcarrier or for a certain subcarrier over multiple time slots is very small, because the channel gains over different subcarriers are different and vary over time slots.

The policy $\mathbb{D}^{L*}$ is a stationary policy. From (3.27) and (3.28), it can be seen that $A_{u,l}^{L*}$ is independent of the time slot. Therefore, $A_{u,l}^{L*}$ which is determined for the $(u, l)$th time slot can be used at the $(u, m)$th time slot $(m = \{0, \ldots, V_L - 1\})$ when states during the two time slots are equivalent (i.e., $\psi_{u,l}^L = \psi_{u,m}^L$). Thus, $\mathbb{D}^{L*}$ is a stationary policy [48].

Since $\mathbb{D}^{L*}$ is a stationary policy, calculating $A_{u,0}^{L*}$ for each state $\psi_{u,0}^L \in \Psi^L$ at the $(u, 0)$th time slot is sufficient to determine $\mathbb{D}^{L*}$. Then, $R_{i,u}^L(\psi_{u,0}^L, A_u^U, \mathbb{D}^{L*})$ and $P_{avg,i}^{C*}, \forall i$ can be found by solving (3.5) and (3.10), respectively. To solve (3.5) and (3.10), the methods explained in [48] can be used with (3.5) and (3.10) written as Bellman optimality equations. Values of $R_{i,u}^L(\psi_{u,0}^L, A_u^U, \mathbb{D}^{L*})$ and $P_{avg,i}^{C*}, \forall i$ are then feedback to the upper-level to update $\lambda$, $\xi$ and $\mu$. The algorithm which determines $\mathbb{D}^{L*}$ is shown in Algorithm 2.

---

**Algorithm 2** : Lower-Level Policy

---

**input**  : $\psi^L, \mathcal{S}_N, \lambda, \xi$ and $\mu$

**output** : $A_{u,l}^{L*} = \{\rho_{i,k}^{C*}, \bar{P}_{i,k}^{C*} | \forall i \in \mathcal{S}_N, k \in \mathcal{K}^C\}$ for each $\psi^{L(2)} \in \Psi^L$, and $R_{i,u}^L(\psi^L, A_u^U, \mathbb{D}^{L*})$ and
$\qquad P_{avg,i}^{C*}, \forall i$

**for** *each* $\psi^{L(2)} \in \Psi^L$ **do**

$\qquad$ Calculate $\bar{P}_{i,k}^{C*}$ and $\rho_{i,k}^{C*}$ by (3.27) and (3.28) at state $\psi^{L(2)}$;

$\qquad r_{i,u,l}^L(\psi^{L(2)}, A_u^U, A_{u,l}^{L*}) \leftarrow \sum_{k \in \mathcal{K}^C} \rho_{i,k}^{C*} R_{i,k}^C(P_{i,k}^{C*})$;

$\qquad P_{tot,i,l}^C(\psi^{L(2)}, A_u^U, A_{u,l}^{L*}) \leftarrow \sum_{k \in \mathcal{K}^C} \bar{P}_{i,k}^{C*}$;

**end**

For each user, calculate $R_{i,u}^L(\psi^L, A_u^U, \mathbb{D}^{L*})$ by solving the Bellman optimality equation written using (3.5);

Calculate $P_{avg,i}^{C*}, \forall i$ by (3.10);

---

The MMDP based resource allocation algorithm, which consists of $\mathbb{D}^{U*}$ and $\mathbb{D}^{L*}$, efficiently allocates resources of the interworking system. However, it has a high time complexity as it requires to find $A_0^{U*}$ and $A_{u,0}^{L*}$ for each system state, and there are total of $(K_S)^{N|\mathcal{K}^C|+|\mathcal{S}_M|}$ system states. The number of system states in this system is significantly high as there are a large number of users and OFDM subcarriers. Therefore, a heuristic resource allocation algorithm with low time complexity is derived in the next section.

## 3.7  Heuristic Resource Allocation

Heuristic resource allocation algorithm consists of two steps. The first step is executed only once at the beginning, and it calculates the dual variables which correspond to data and voice traffic constraints (i.e., $\lambda$ and $\xi$) based on the average square channel gains ($\Omega$'s). That is, $\Omega = \mathbb{E}\{h^2\}$, where $\mathbb{E}\{\cdot\}$ is the ensemble average operator and $h$ is the channel gain. The second step uses the dual variables calculated in the first step, and allocates upper and lower level resources based on the instantaneous channel gains subject to total power constraints of the users (i.e., C4).

A key difference between the heuristic and the MMDP based algorithms is as follows. The heuristic algorithm executes the resource allocation algorithm (to be precise, the second step of the algorithm) at each time slot to determine resource allocation decisions. On the other hand, the MMDP based algorithm determines upper and lower level policies which consist of sets of resource allocation decisions for each possible system state, at the beginning. After the initial policy calculation, at each time slot the policies of the MMDP based algorithm simply map the system state to a set of resource allocation decisions. However, when the system changes, recalculation of the policies is required.

The heuristic algorithm has a low time complexity due to two reasons. First, as each step of the algorithm is executed based on a single system state, which consists of $\Omega$'s for the first step and instantaneous channel gains for the second step, solving Bellman optimality equations is not required to find $R_i^U(\psi_{0,0}, \mathbb{D})$, $R_{i,u}^L(\psi_{u,0}^L, A_u^U, \mathbb{D}^L)$ and $P_{avg,i}^C$. In the MMDP based algorithm, the Bellman optimality equations are solved by determining $A_0^U$ and $A_{u,0}^L$ for each possible system state. Second, users are allocated for contention-based channel access of the WLAN using a simple algorithm (to be discussed).

The first step finds $\lambda$ and $\xi$ by solving $\mathcal{P}2$, $\mathcal{P}3$ and $\mathcal{P}4$. Therefore, the solutions obtained for $\mathcal{P}3$ and $\mathcal{P}4$ in Sects. 3.5.2 and 3.6 are used in this step with modifications to utilize $\Omega$'s as follows. Average throughput over a Rayleigh fading channel is given by [45, 55]

$$\mathbb{E}\{R\} = \int_0^\infty \frac{2Bh}{\Omega} \log_2(1 + \frac{h^2 p}{n}) e^{-\frac{h^2}{\Omega}} \, dh = \frac{B}{\ln(2)} e^{\frac{n}{\Omega p}} \mathrm{E}_1\left(\frac{n}{\Omega p}\right),\qquad (3.29)$$

where $p$ is the transmit power level, $B$ is the bandwidth, $n$ is the total noise plus interference power, and $\mathrm{E}_1(x)$ is the exponential integral which is defined as [55]

$$\mathrm{E}_1(x) = \int_x^\infty \frac{e^{-x}}{x} \, dx. \qquad (3.30)$$

As $0.5e^{-x}\ln(1 + 2x)$ provides a tight lower bound for $\mathrm{E}_1(x)$ [55], from (3.29)

$$\mathbb{E}\{R\} > \frac{B}{2} \log_2\left(1 + \frac{2\Omega p}{n}\right). \qquad (3.31)$$

Therefore, the solutions for $\mathcal{P}3$ and $\mathcal{P}4$ are modified to calculate throughput over each wireless channel by $(B/2)\log_2(1 + 2\Omega p/n)$. That is, the equations in Sects. 3.5.2 and 3.6 are modified by substituting $B/2$ to $B$, and $2\Omega$ to $h^2$. The latter is also equivalent to the substitution of $2\mathbb{E}\{\alpha_{i,y}^n\}$ to $\alpha_{i,y}^n$. Since $\Omega$'s are used in this step, there is only one possible state in the MMDP. Consequently,

$$R_i^U(\psi_{0,0}, \mathbb{D}) = r_{i,0}^U(\psi_{0,0}, A_0^U, \mathbb{D}^L), \qquad (3.32)$$

$$R_{i,0}^L(\psi_{0,0}^L, A_0^U, \mathbb{D}^L) = r_{i,0,0}^L(\psi_{0,0}^L, A_0^U, A_{0,0}^L) \qquad (3.33)$$

and

$$P_{avg,i}^C = \sum_{k \in \mathcal{K}^C} \rho_{i,k}^C P_{i,k}^C. \qquad (3.34)$$

The first step of the heuristic algorithm is shown in Algorithm 3.

---

**Algorithm 3** : First Step of the Heuristic Algorithm

---

**input**    : Average square channel gains ($\Omega$'s), $\mathcal{S}_M$, $\mathcal{S}_S$, and $P_{T,i}$, $R_{Vmin,i}$ and $R_{Dmin,i}$, $\forall i$

**output**   : $\lambda^*, \xi^*$ and $\mathcal{S}^{CB*}$

Form $\psi_0^U$ and $\psi_{0,0}^L$ using $\sqrt{2\Omega}$ values of the channels;

$\alpha_{i,k}^C \leftarrow 2\mathbb{E}\{\alpha_{i,k}^C\}$ and $\alpha_i^W \leftarrow 2\mathbb{E}\{\alpha_i^W\}$;

Sort users in $\mathcal{S}_M$ in the descending order of their $\mathbb{E}\{\alpha_i^W\}$;

**while** $\mathcal{S}^{CB}$ *corresponding to maximum* $\sum_{i \in \mathcal{S}_N} R_i^U(\psi_{0,0})$ *is not found* **do**

    $\mathcal{S}^{CB} \leftarrow \mathcal{S}_M(|\mathcal{S}^{CB}| + 1)$;

    $\lambda \leftarrow \{0, \ldots, 0\}, \xi \leftarrow \{0, \ldots, 0\}$ and $\mu \leftarrow \{\mu_1, \ldots, \mu_N\}$;

    **while** $\lambda$ *and* $\xi$ *are not optimal* **do**

        **while** $\mu$ *is not optimal* **do**

            Calculate $\bar{P}_{i,j}^{CF*}$, $\rho_{i,j}^{CF*}$, $P_i^{CB*}$, $\rho_{i,k}^{C*}$ and $\bar{P}_{i,k}^{C*}$ by (3.18)–(3.21), (3.23), (3.24), (3.27) and

            (3.28) substituting $B^W/2$ to $B^W$ and $\Delta f/2$ to $\Delta f$;

            Update $\mu_i$, $\forall i$;

        **end**

        $R_{i,0}^L(\psi_{0,0}^L) \leftarrow \sum_{k \in \mathcal{K}^C} \bar{R}_{i,k}^C(\bar{P}_{i,k}^C, \rho_{i,k}^C)$;

        $P_{avg,i}^C \leftarrow \sum_{k \in \mathcal{K}^C} \bar{P}_{i,k}^C$;

        Calculate $r_{i,0}^U(\psi_{0,0})$ by (3.6), and $R_i^U(\psi_{0,0}) \leftarrow r_{i,0}^U(\psi_{0,0})$;

        Update $\lambda_i$ and $\xi_i$, $\forall i$;

    **end**

**end**

---

In the second step, upper and lower level resources are jointly allocated at the beginning of the $u$th time slot ($u = \{0, 1, 2, \ldots\}$) subject to C4 and assuming that the current lower-level state remains unchanged within the $u$th time slot (i.e., $\psi_{u,0}^L = \psi_{u,l}^L$, $\forall l \in \{1, \ldots, V_L - 1\}$). With this assumption,

$$R_{i,u}^L(\psi_{u,0}^L, A_u^U, \mathbb{D}^L) = r_{i,u,0}^L(\psi_{u,0}^L, A_u^U, A_{u,0}^L) \tag{3.35}$$

and

$$P_{avg,i}^C = \sum_{k \in \mathcal{K}^C} \rho_{i,k}^C P_{i,k}^C. \tag{3.36}$$

Resources are allocated using the algorithms which solve $\mathcal{P}3$ and $\mathcal{P}4$. In this step, these algorithms only calculate $\mu$ while using $\lambda$ and $\xi$ from the first step. At the beginning of the $u$th time slot, they distribute UE power between WLAN and cellular network interfaces. At the beginnings of the remaining $(u, l)$th time slots (i.e., $l = \{1, \ldots, V_L - 1\}$), the subcarriers and the amount of power dedicated for the cellular network interfaces are reallocated based on the current state $\psi_{u,l}^L$ to fully exploit the multi-user diversity in the cellular network. The second step of the heuristic algorithm is shown in Algorithm 4.

---

**Algorithm 4** : Second Step of the Heuristic Algorithm

---

**input** : Instantaneous channel gains during $(u, l)$th time slot, $\lambda^*, \xi^*$ and $\mathcal{S}^{CB*}$

**output** : $A^{U*} = \{P_i^{CB*}, \rho_{i,j}^{CF*}, \bar{P}_{i,j}^{CF*} | \forall i \in \mathcal{S}_M, j \in \mathcal{K}^{CF}\}$ and $A^{L*} = \{\rho_{i,k}^{C*}$ and $\bar{P}_{i,k}^{C*} | \forall i \in \mathcal{S}_N, k \in \mathcal{K}^C\}$

Form $\psi_u^U$ and $\psi_{u,l}^L$ using instantaneous channel gains during $(u, l)$th time slot;

**while** $\mu$ *is not optimal* **do**

    **if** $l = 0$ **then**

        Calculate $\bar{P}_{i,j}^{CF*}, \rho_{i,j}^{CF*}, P_i^{CB*}, \rho_{i,k}^{C*}$ and $\bar{P}_{i,k}^{C*}$ by (3.18)–(3.21), (3.23) and (3.24), (3.27) and (3.28) based on $\psi_u^U$ and $\psi_{u,l}^L$;

    **else**

        Recalculate $\rho_{i,k}^{C*}$ and $\bar{P}_{i,k}^{C*}$ by (3.27) and (3.28) based on $\psi_{u,l}^L$;

    **end**

    Update $\mu_i, \forall i$;

**end**

---

Users are allocated for contention-based channel access of the WLAN as follows. The optimal set of users who use contention-based channel access consists of a few users with strong channel conditions due to two reasons. First, allocating too many users for contention-based channel access degrades the aggregated throughput of the users as more collisions occur in the wireless channel [12]. Second, allocating a user with a weaker channel degrades throughputs of all the users as the weak user takes a longer time to transmit a packet [15]. Therefore, users for the contention-based channel access are allocated as follows. First, users in $\mathcal{S}_M$ are sorted in the descending order of their $\mathbb{E}\{\alpha_i^W\}$. Second, the first step of the heuristic algorithm is repeated, each time adding the next user in $\mathcal{S}_M$ to $\mathcal{S}^{CB}$, until the total throughput achieved at the upper-level reaches the maximum.

## 3.8 Performance Evaluation

Performance of the proposed resource allocation schemes are evaluated in a system which consists of a cellular network and a WLAN as shown in Fig. 3.1. Wireless channels in the system are modeled as Rayleigh fading channels, and their path loss is proportional to $d^{-4}$, where $d$ denotes the distance between users and the WLAN AP or the cellular BS. These channels over the WLAN are generated at carrier frequency of 2.4 GHz and mobile speed of $3 \, \text{km h}^{-1}$, while those over the cellular network are generated at carrier frequency of 2.1 GHz and mobile speed of $50 \, \text{km h}^{-1}$. Based on the channel coherence times, $T^L$ and $T^U$ are selected to be 4.23 ms and 63.45 ms, respectively [56, 57]. Radiuses of the WLAN and the cellular coverage areas are 50 m and 1000 m, respectively. Users are uniformly distributed over the coverage areas. Total power available at each user is uniformly distributed between 0 and 1 W. Table 3.1 shows the remaining parameters.

**Table 3.1** Simulation parameters

| Parameter | Value (unit) |
|---|---|
| $B^W$ | 20 MHz |
| $D$ | 4095 octets |
| $|\mathcal{K}^C|$ | 4 or 128 |
| $|\mathcal{K}^{CF}|$ | 2 or 10 |
| $T^L$ | 4.23 ms |
| $T^U, T_P$ | 63.45 ms |
| $T_{ACK}$ | 24.5 μs |
| $T_{AIFS}$ | 34 μs |
| $T_{CF}$ | $31.72/|\mathcal{K}^{CF}|$ ms |
| $T_{CP}$ | 31.72 ms |
| $T_{CTS}$ | 24.5 μs |
| $T_{RTS}$ | 24.7 μs |
| $T_{SIFS}$ | 16 μs |
| $\Delta f$ | $5/|\mathcal{K}^C|$ MHz |
| $\sigma_0$ | 9 μs |
| Additive white Gaussian noise density | −174 dBm/Hz |
| Initial window size of WLAN | 16 |
| Maximum number of backoff stages in WLAN | 6 |

First, performance of the MMDP based resource allocation algorithm (MM) and the heuristic algorithm (HM) is evaluated in a small-scale system, denoted by system-1, and compare the performance with that of a benchmark algorithm (BM1). Second, performance of HM is evaluated in a large-scale system. BM1 resembles the first category resource allocation algorithms described in Sect. 3.2, and it allocates resources as follows. First, it assigns users for the two networks via exhaustive search such that the total average system throughput is maximized. In this step, average users' throughputs are calculated using $(B/2) \log_2(1 + 2\Omega p/n)$, similar to the first step of HM. Second, each network individually allocates its resources at the every time slot to maximize the network throughput, using instantaneous channel gains. BM1 allocates resources at two time-scales based on the PHY and MAC technologies of the networks, but it does not allow UE multi-homing.

In system-1, there are four users ($|\mathcal{S}_S| = |\mathcal{S}_M| = 2$), four subcarriers and two contention-free TXOPs. Two-state Markov channels are used [58]. The boundary between the two states of each channel is selected such that the steady state probability of each state is 0.5. For each channel, the channel is at the first state if $h < \sqrt{\Omega \ln(2)}$; otherwise, it is at the second state. When a channel is at the first and the second states, channel gains are $\sqrt{\Omega(1 - \ln(2))}$ and $\sqrt{\Omega(1 + \ln(2))}$, respectively. These channel gains are calculated by averaging the power gain of a

continuous-envelope Rayleigh fading channel within boundaries of the corresponding state. Transition probabilities between the states are calculated as described in
[45]. In MM, $\theta$ and $\beta$ are set to be 0.9.

Throughputs achieved by MM, HM and BM1 for different QoS requirements
in system-1 are compared in Fig. 3.4. MM provides throughput improvement of at
least 10.7% compared to HM. Both MM and HM provide higher throughputs than
BM1 as they enable multi-homing, whereas BM1 allows each user to access only
one network. When multi-homing is enabled, users achieve higher throughputs due
to efficient resource utilization, which is a result of jointly allocating resources of
multiple networks and catering for user QoS requirements using multiple network
resources. MM outperforms HM as MM allocates resources statistically considering
the future state changes using an MMDP, whereas HM allocates resources based on
the current state only. For example, in the second step of HM, resources of the two
networks are jointly allocated at the beginning of each slow time-scale time slot.
Transmit power of each UE is also distributed among the two network interfaces of
the UE. Now, even if the channel conditions change over the fast time-scale time

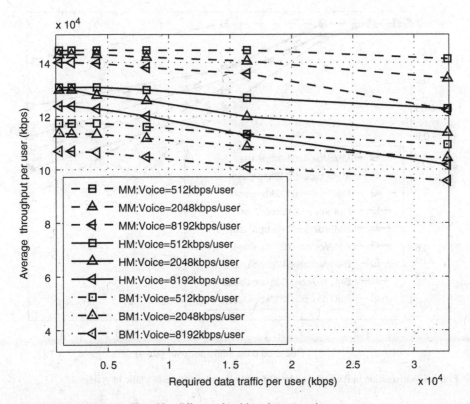

**Fig. 3.4** Throughputs achieved by different algorithms in system-1

slots, the transmit power distribution remains the same. Thus, HM provides less throughput performance compared to MM.

Users' data traffic requirement satisfaction performance of the algorithms is shown in Fig. 3.5. It is measured using satisfaction index (SI)[52], which is defined for a particular traffic class as

$$SI = \mathbb{E}\left\{ 1_{R \geq R_{min}} + 1_{R_{min} > R} \cdot \frac{R}{R_{min}} \right\}, \qquad (3.37)$$

where $R$ and $R_{min}$ are achieved and required throughputs respectively; and $1_{x \geq y} = 1$ if $x \geq y$ or $1_{x \geq y} = 0$ otherwise. Similar to throughput performance of these algorithms, MM and HM achieve higher SI's for data traffic ($SI_D$) compared to BM1, providing users with better QoS. The difference between these $SI_D$'s is significant at high data traffic requirements as multi-homing is particularly useful for catering for higher user QoS requirements via multiple networks. All three algorithms achieve SI for voice traffic ($SI_V$) of one in system-1.

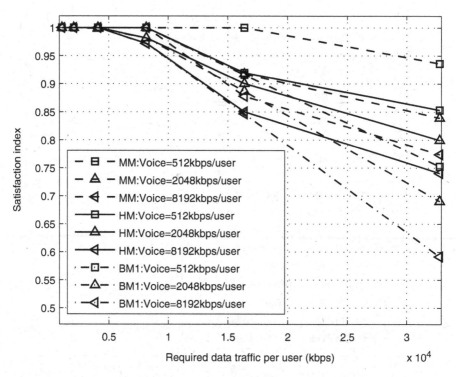

**Fig. 3.5** Satisfaction index achieved by different algorithms for data traffic in system-1

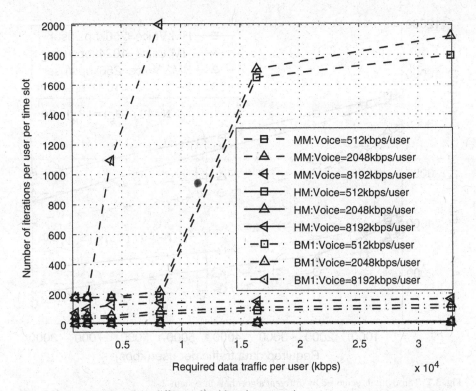

**Fig. 3.6** Complexities of the different algorithms in system-1

Complexities of the algorithms in system-1 are shown in Fig. 3.6. Complexities of the algorithms are measured in terms of the required number of iterations in the inner most loop per user per fast time-scale time slot. MM requires a large number of iterations as it solves an MMDP based resource allocation problem with $2^{18}$ system states. BM1 requires a higher number of iterations than HM requires, as BM1 recalculates $\lambda$ and $\xi$ at each time slot, whereas HM calculates $\lambda$ and $\xi$ only once in its first step.

Performance of HM is also evaluated in a large-scale system, denoted by system-2, and is compared with performance of a benchmark algorithm (BM2). Performance of MM is not evaluated in this system due to MM's high complexity. System-2 consists of 128 subcarriers, 10 contention-free TXOPs and 40 or 80 users. One hundred twenty-eight subcarriers are used in this system as the highest number of resource blocks per cell in a LTE system is 110. Wireless channels in the system are modeled as continuous envelope Rayleigh fading channels generated at the same

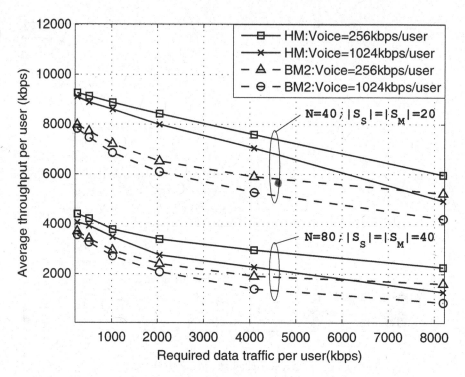

**Fig. 3.7**  Throughputs achieved by different algorithms in system-2

carrier frequencies and mobile speeds as in system-1. BM2 uses a simpler user allocation mechanism than exhaustive search used in BM1, as exhaustive search is not feasible in system-2 due to presence of a large number of users. It allocates users to the two networks as follows. First, it sorts users in $\mathcal{S}_M$ in the descending order of users' $\mathbb{E}\{\alpha_i^W\}$. Second, it individually allocates resources of the networks $|\mathcal{S}_M|$ times while calculating users' average throughputs using $(B/2)\log_2(1 + 2\Omega p/n)$. At the $j$th resource allocation, it allocates the first $j$ users in $\mathcal{S}_M$ to the WLAN, and the remaining users in $\mathcal{S}_M$ and all the users in $\mathcal{S}_S$ to the cellular network. Finally, the user allocation which resulted in the highest aggregated average throughput is selected. After the user allocation, BM2 allocates resources of the networks at each time slot, similar to BM1.

Throughput, $\mathrm{SI}_V$ and $\mathrm{SI}_D$ performance of HM and BM2 in system-2 are shown in Figs. 3.7, 3.8 and 3.9, respectively. HM provides higher throughput, $\mathrm{SI}_V$ and $\mathrm{SI}_D$ performance than BM2 as it enables user multi-homing. Performance of both algorithms decreases with the number of users, as resources of the two networks are distributed among more users when each user has a certain QoS requirement.

Complexities of HM and BM2 in system-2 are shown in Fig. 3.10. HM converges within 25 iterations per user per time slot while BM2 requires more than 47 iterations per user per time slot to converge. Complexity of HM does not significantly

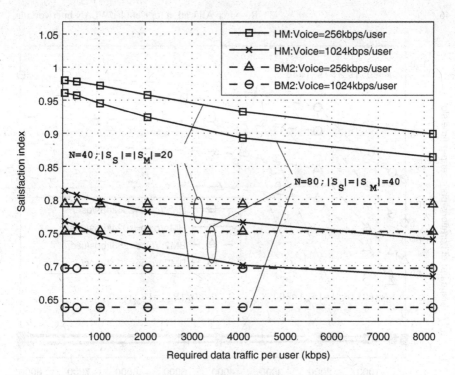

**Fig. 3.8** Satisfaction index achieved by different algorithms for voice traffic in system-2

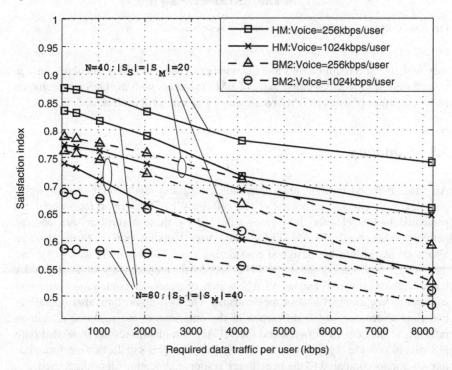

**Fig. 3.9** Satisfaction index achieved by different algorithms for data traffic in system-2

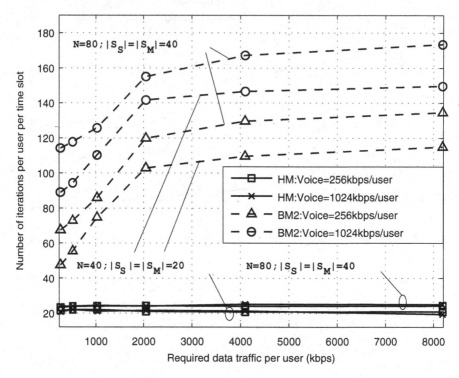

**Fig. 3.10** Complexities of different algorithms in system-2

vary with the QoS requirements and the number of users as HM recalculates only $\mu$ at each time slot, whereas complexity of BM2 increases with the QoS requirements and the number of users as BM2 recalculates $\lambda$, $\xi$ and $\mu$ at each time slot.

## 3.9   Summary

An MMDP based optimal algorithm and a heuristic algorithm with low time-complexity to allocate resources of the cellular/WLAN interworking system are presented in this chapter. Resources allocated by these algorithms are cellular network subcarriers, WLAN TXOPs and user transmit power. Both resource allocation algorithms are designed considering multi-homing users with voice and data traffic requirements and the PHY and MAC technologies of an OFDMA based cellular network and a WLAN which operates on contention-based and contention-free channel access mechanisms. The two algorithms also operate at two time-scales due to the difference in the durations of the resource allocation intervals of the cellular network and the WLAN. Simulation results show that both algorithms provide significant user throughput and QoS satisfaction performance improvements compared to the benchmark resource allocation algorithms used.

# Chapter 4
# Resource Allocation for D2D Communication Underlaying Cellular/WLAN Interworking

This chapter investigates uplink resource allocation for D2D communication underlaying cellular network and WLAN interworking system. D2D communication is enabled in this interworking system as it provides two main advantages: (1) it allows setting up secure high capacity WLAN based D2D links by sending control signals and authentication related information through the cellular network using user multi-homing capability; and (2) it enhances network throughput by incorporating hop and reuse gains to the network. The key challenges for allocating resources for this system are selecting users' communication modes to maximize hop and reuse gains, interference management and allocating resources based on PHY and MAC technologies of different networks. These challenges and the solutions to overcome them are also investigated in this chapter.

This chapter first discusses the challenges for allocating resources in the system, and explains existing and new solutions. Second, it presents the D2D communication underlaying cellular/WLAN interworking system model. Third, a resource allocation scheme is derived addressing those challenges, and its implementation is discussed. Fourth, performance of the derived scheme is evaluated via simulations.

## 4.1 Challenges for Resource Allocation

There are three main technical challenges for allocating resources in a D2D communication underlaying cellular/WLAN interworking system: (1) allocation of resources based on multiple radio access technologies, (2) selection of users' communication modes for multiple networks to maximize hop and reuse gains, and (3) interference management. These challenges and the solutions to them are discussed in the following sections.

© The Author(s) 2018     47
A.T. Gamage, X.(S.) Shen, *Resource Management for Heterogeneous Wireless Networks*, SpringerBriefs in Electrical and Computer Engineering,
DOI 10.1007/978-3-319-64268-0_4

### 4.1.1  Challenge 1: Multiple Radio Access Technologies

The challenge for allocating resources based on multiple PHY and MAC technologies in an interworking system is discussed in detail in Chap. 3. The first proposed solution in Chap. 3 to overcome this challenge is to accurately estimate user throughputs and transmit power consumptions based on PHY and MAC of the cellular network and the WLAN as follows. The user throughputs over the OFDMA based cellular network is estimated using the Shannon capacity formula. The user throughputs and the transmit power consumptions for the WLAN are estimated as average achievable throughputs and average power consumptions calculated considering the effect of collisions in the wireless channel. The second proposed solution is to operate the resource allocation algorithm in two time-scales as the resource allocation intervals of these two networks are different. To model the resource allocation problem over two time-scales, a MMDP based optimal approach and a simple heuristic approach are discussed in Chap. 3.

### 4.1.2  Challenge 2: Efficient Mode Selection

Selecting the best communication modes, which take advantage of user proximity and fully realize hop and reuse gains, is crucial to achieve higher data rates with enhanced QoS for the users. There are two challenges for efficient mode selection: (1) high complexity and communication overhead, and (2) realization of hop and reuse gains.

Mode selection process has a high complexity as it requires estimation of a large number of channels due to availability of a large number of potential D2D and traditional links over multiple networks in the interworking system. It also causes a large communication overhead due to transmission of a large volume of CSI [18]. Thus, repeating mode selection in a very fast time-scale, such as at every resource allocation interval in the cellular network, to calculate the best communication modes based on instantaneous CSI is not practical. To reduce the complexity and the communication overhead, mode selection can be performed in a slower time-scale based on channel statistics. This time-scale should not be too slow as well, because the D2D links may become weak over time due to user mobility. This issue can be relaxed in the interworking system by creating D2D links for high mobility users via the cellular network only, as the cellular communication has a longer range. D2D links for low mobility users can be created via both the networks.

Selecting communication modes to maximize hop gain is a challenge. To estimate the hop gain achieved by selecting the D2D mode for a certain pair of users, total throughput achieved via the D2D mode and throughput achieved if those users were to use the traditional mode should be calculated. The total throughput via the D2D mode is the sum of D2D link throughput and additional throughput achieved utilizing saved resources. Resources are saved when D2D mode is used, as D2D links use either UL or DL resources only. The additional throughput will be in the UL if the DL resources are used for the D2D links, and vice versa. In a time-division duplexing (TDD) system, throughput of the D2D mode can be calculated by allocating all the resources to the D2D link, and throughput of the traditional mode can be calculated by allocating a part of the resources to the UL and remaining resources to the DL [59]. This method provides accurate results as UL and DL share the same set of resources in a TDD system. However, in frequency-division duplexing (FDD) systems, joint allocation of UL and DL resources is necessary as UL and DL use two dedicated sets of resources over two different carrier frequencies. As calculation of the total throughput of D2D mode is complicated, selecting the communication modes to maximize hop gain is a challenge.

Selecting communication modes to achieve a high reuse gain in a network is challenge due to two reasons: (1) finding the optimal pair of D2D and traditional links to reuse (i.e., share) a particular resource is tedious as there are a large number of different link pairs to be considered for each resource; and (2) calculation of optimal power levels for D2D and traditional links over a reused resource is complicated as there is CCI between the links. In [59–61], power allocation to capture reuse gain is investigated. These works assume that the number of available resources equals to the number of traditionally communicating users, and that each traditionally communicating user occupies only one resource. Power allocation to maximize the D2D link throughput when this D2D link reuses all the resources is studied in [61]. When there are multiple D2D links and each D2D link reuses only one resource, the optimum power allocation to maximize the total throughput over a resource is found in [59, 60]. However, in large multicarrier systems (e.g., LTE-A networks), there are a large number of subcarriers or physical resource blocks (RBs) compared to the number of users. Furthermore, the set of resources allocated to one traditional link could be reused by several D2D links, where each D2D link reuses a subset of the resources allocated to the traditional link, and vice versa. In this scenario, allocation of cellular network RBs based on a reverse iterative combinatorial auction based algorithm is investigated in [62, 63].

### 4.1.3 Challenge 3: Interference Management

In D2D communication underlaying interworking systems, CCI and intercarrier interference (ICI) caused by D2D communication degrade throughput performance of the system. ICI occurs in multicarrier systems, such as LTE-A networks, due to signals transmitted by different users over different subcarriers arrive at a receiver with different delays [64]. Managing this interference to maximize system throughput is a challenge, as it adds more complexity to the resource allocation process by requiring to make three additional decisions: (1) whether to allocate orthogonal or non-orthogonal resources for D2D links, (2) whether to utilize DL or UL resources for D2D links, and (3) select D2D and traditional link pairs to reuse resources, and determine transmit power levels for these link pairs as discussed under *Challenge 2*. Furthermore, to ease interference management, characteristics of the interworking systems, such as low transmit power levels of multi-homing users, should be taken into account during the resource allocation. Transmit power levels of multi-homing users tend to be lower than that of users communicating via only one network, as their transmit power is distributed over two or more networks.

Orthogonal or non-orthogonal resources for a particular D2D link should be chosen based on throughputs achieved using each of these resource types. An orthogonal resource is used by only one link whereas a non-orthogonal resource is reused by a D2D and a traditional link. Therefore, when non-orthogonal resources are used, there is CCI between the links. When these two links are in different proximities, total throughput of the two links is shown in Fig. 4.1, where $P_t$ and $P_d$ are transmit power levels of traditional and D2D link transmitters, and the links reuse a LTE-A network RB. As shown in Fig. 4.1a, allocating non-orthogonal resources for a D2D link is advantageous when the D2D link is far away from the traditional link and the two D2D communicating users are in proximity, due to limited CCI between the links [59, 60]. When the two links are in a close range, a higher total throughput can be achieved by allocating orthogonal resources as shown in Fig. 4.1b, c. The total throughput in this scenario reaches the highest when the link with higher channel gain uses the RB.

Selection of UL or DL resources for D2D links affect differently the interference management and system complexity. When DL resources are reused for D2D links, traditionally communicating users will suffer from CCI. To calculate power levels of D2D link transmitters ensuring tolerable CCI at traditionally communicating users, estimating channels between D2D link transmitters and traditionally communicating users is required. Moreover, this CCI could be severe if a D2D pair and a traditionally communicating user are located at a cell edge or at nearby cell edges [18]. On the other hand, when UL resources are reused for D2D links, BSs will suffer from CCI. To manage this CCI, already available CSI of the channels between users and BSs can be utilized. In addition to CCI, cellular network users will suffer from ICI when DL resources are utilized for D2D links, as signals from the BS and D2D transmitters arrive at the users at different time instances. If UL resources are utilized for D2D links, BSs will suffer from ICI. However, unlike users, BSs are equipped with ICI cancellation schemes as ICI is an inherent problem in OFDMA based UL systems, such as LTE-A networks. Therefore, use of UL resources for D2D links simplifies CCI and ICI management. Furthermore, use of UL resources for D2D links is beneficial as UL resources are less utilized compared to DL resources due to asymmetric UL and DL traffic loads [60, 61].

Interworking of networks simplifies interference management in three ways. First, low transmit power levels of multi-homing users in interworking systems cause low CCI. Second, CCI can be reduced by selecting the resources to be reused from a large number of resources, which are available from multiple networks, with different channel conditions. Third, interworking of cellular networks and WLANs enables setting up of WLAN based D2D links [65]. Multiple of these D2D links can be setup and simultaneously operated among users in proximity without causing

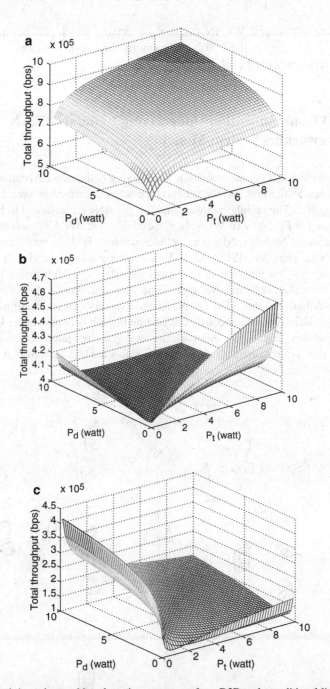

**Fig. 4.1** Total throughput achieved reusing a resource for a D2D and a traditional link. (**a**) D2D and traditional links are far away from each other. (**b**) Two links are in proximity, and the traditional link has a higher channel gain. (**c**) Two links are in proximity, and the D2D link has a higher channel gain

CCI, as there are several WLAN frequency channels which can be utilized for these links. For example, IEEE 802.11n supports three non-overlapping channels in 2.4 GHz frequency band [12].

## 4.2  D2D Communication Underlaying Cellular/WLAN Interworking System Model

System model for D2D communication underlaying cellular/WLAN interworking system focuses on the first and the second type areas described in Sect. 2.1, and is shown in Fig. 4.2. The cellular network and the WLANs are assumed to be a LTE-A network and IEEE 802.11n WLANs, respectively. WLAN APs' and the LTE-A network enhanced NodeB (eNB) (i.e., cellular network BS) are interconnected via Internet service provider (ISP) and LTE-A evolve packet core (EPC) network. APs are synchronized with the LTE-A network using synchronization protocols, such as IEEE 1588-2008 and Network Time Protocol version 4 (NTPv4), over Ethernet backhauls of the APs. Users can access services connecting to either network, e.g., $UE_8$, or simultaneously connecting to both networks using UE multi-homing capability, e.g., $UE_9$. Network assisted (i.e., operator controlled) D2D communication is considered. In addition to setup a D2D link over a network,

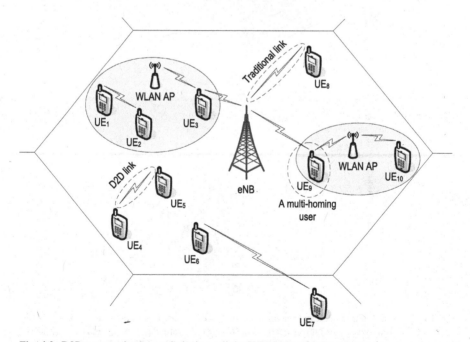

**Fig. 4.2** D2D communication underlaying cellular/WLAN interworking system

multiple D2D links can also be setup for the same pair of users using the UE multi-homing capability. For example, two D2D links can be setup between $UE_1$ and $UE_2$ over the LTE-A network and the WLAN. When two users in proximity are not within an AP coverage, a high capacity D2D link can be setup between them by pairing their WLAN radios. To pair the WLAN radios, relevant control information and authentication request/response messages are sent through the LTE-A network. When a user access services that cannot be accessed via D2D links, such as accessing Internet, email and voice mail, user's communication is referred to as non-D2D communication. Each user is assumed to require either non-D2D or D2D communication.

## 4.3  Three-Time-Scale Resource Allocation

In this section, a resource allocation scheme for D2D communication underlaying cellular/WLAN interworking system is derived overcoming the challenges discussed in Sect. 4.1. The key design objectives are: (1) maximize the total system throughput subject to user QoS and total power constraints; and (2) minimize the signaling overhead and the computational complexity such that this scheme can be employed in practical systems. The total system throughput is the sum of all the D2D and traditional link throughputs achieved over both networks. Then, the resource allocation problem for this system can be stated as follows.

$$\mathcal{P}6 : \max \quad \sum_{i \in \mathcal{S}_N} \left( R_i^{C(D)} + R_i^{C(T)} + R_i^{W(D)} + R_i^{W(T)} \right)$$

$$\text{s.t.} \quad C9 : R_i^{C(T)} + R_i^{W(T)} \geq R_{min,i}^{ND}, \ \forall i \in \mathcal{S}_N$$

$$C10 : R_i^{C(D)} + R_i^{C(T)} + R_i^{W(D)} + R_i^{W(T)} \geq R_{min,i}^{D2D}, \ \forall i \in \mathcal{S}_N$$

$$C11 : P_i^C + P_i^W \leq P_{T,i}, \ \forall i \in \mathcal{S}_N,$$

where $\mathcal{S}_N$ denotes the set of all the users; $R_i^{C(D)}$ and $R_i^{C(T)}$ are the throughputs achieved by the $i$th user through the cellular network using D2D and traditional modes respectively; $R_i^{W(D)}$ and $R_i^{W(T)}$ are the throughputs achieved by the $i$th user through both channel access mechanisms of the WLAN using D2D and traditional modes respectively; $R_{min,i}^{ND}$ and $R_{min,i}^{D2D}$ are the minimum data rates required for the $i$th user's non-D2D and D2D communications respectively; and $P_i^C$ and $P_i^W$ are the total power used by the $i$th user for communications via the cellular network and the two channel access mechanisms of the WLAN respectively. In $\mathcal{P}6$, C9 and C10 are the QoS constraints, and C11 is the total power constraint. If the $i$th user requires only non-D2D communication, then only C9 is active for the user. Otherwise, if the user requires only D2D communication, only C10 is active.

**a**

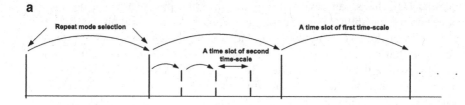

**b**

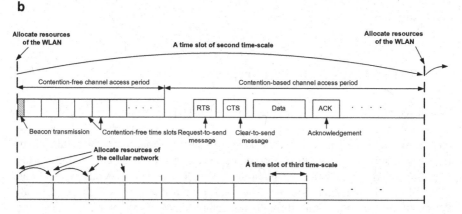

**Fig. 4.3** Proposed three time-scale resource allocation scheme. (**a**) Operations of first and second time-scales. (**b**) Detailed view of a time slot of second time-scale

As shown in Fig. 4.3, the derived resource allocation scheme operates on three different time-scales. The first time-scale is the slowest while the third time-scale is the fastest. That is, a time slot in the first time-scale is the longest while that in the third time-scale is the shortest. Mode selection is performed in the first time-scale. Resources of the cellular network and the WLANs are jointly allocated in the second time-scale. As the cellular network has a very short resource allocation interval compared to that of the WLANs, resources of the cellular network are reallocated in the third time-scale. The derived scheme addresses the challenges stated in Sect. 4.1 as follows.

- To address *Challenge 1*: based on the insights gain from Chap. 3, a low complex joint resource allocation for the cellular network and the WLANs is performed over the second and the third time-scales based on average channel gains and instantaneous CSI; and efficient and feasible resource allocation decisions are made considering PHY and MAC technologies of the two networks.
- To address *Challenge 2*: complexity and signaling overhead are reduced by performing mode selection in the first time-scale; hop gain is evaluated in the mode selection process by using two resources, which can be allocated to a D2D link or a traditional link with a UL and a DL, to calculate the throughput of each

mode; and reuse gain is maximized by allocating non-orthogonal resources via a simplified two-step resource allocation process.

- To address *Challenge 3*: non-interfering WLAN based D2D links are used; CCI and ICI mitigation is simplified by using UL resources for D2D communication within the cellular network; severe CCI is prevented by avoiding allocation of non-orthogonal resources for the links in proximity; and CCI is further reduced by enabling UE multi-homing for both D2D and traditional modes, and properly calculating the UE transmit power.

## 4.3.1 First Time-Scale: Mode Selection

Mode selection is performed in the first time-scale to reduce the average computational complexity and the average signaling overhead by less frequent mode selection decisions and corresponding channel estimations. Furthermore, mode for each user over each network is selected separately as wireless channel gains over different networks are different.

Mode selection algorithm is shown in Fig. 4.4, where CCS is a centralized control server attached to the EPC. As shown in *Step 1*, the first step of the mode selection algorithm is to determine communication modes for the users within an AP coverage. These users can access both networks using the UE multi-homing capability. In *Step 2*, if a pair of users is not within an AP coverage, but within WLAN radio communication range, the pair is allocated a WLAN based D2D link. For these links, $T_{CFP} = T_P$ and the pair of users uses all the available TXOPs. Throughput over each TXOP is given by (2.2). Due to high capacity of these links, when a user pair is allocated one of these links, these users are not allocated cellular network resources for D2D communication. In *Step 2*, when the pair of users are not within the WLAN radio communication range, the users are allocated cellular network resources.

In the cellular network, mode for each user is selected based on the achievable throughputs using each mode, utilizing two RBs. In the traditional mode, throughput is calculated allocating one RB for UL and the other for DL. In the D2D mode, throughput is calculated allocating both RBs for the D2D link to capture the hop gain. Moreover, throughputs are calculated substituting average channel gains and unit transmit power levels into (2.1), due to following two reasons. There are multiple third time-scale time slots within one of the first time-scale time slot. Thus, the transmit power levels vary within a first time-scale time slot. As the duration of a first time-scale time slot is long, instantaneous channel gains change within a first time-scale time slot. In the WLANs, mode selection is performed in a similar manner, using two contention-free TXOPs. In each mode, throughput over a TXOP is calculated using (2.2). Users use the same mode for both contention-based and contention-free channel access mechanisms.

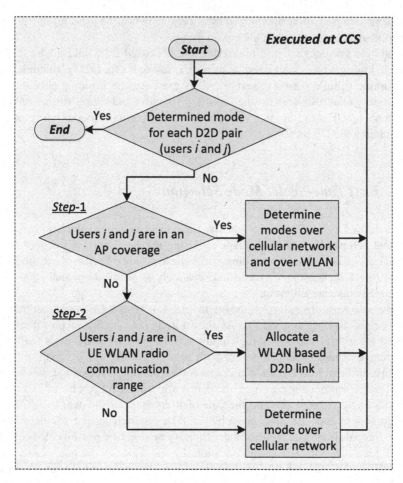

**Fig. 4.4** Mode selection algorithm

### 4.3.2 Second Time-Scale: Joint Resource Allocation for Cellular Network and WLANs

Second time-scale resource allocation jointly allocates cellular network and WLAN resources, and transmit power available at multi-homing users, ensuring QoS satisfaction. These resources are allocated using average channel gains over the cellular network and instantaneous channel gains over the WLANs. Average channel gains over the cellular network are used as there are multiple third time-scale time slots within one time-slot of the second time-scale. Instantaneous channel gains over the WLANs are used as the second time-scale time slot duration is shorter than the coherence time of the channels over WLANs (see Sects. 3.1 and 3.3.1). Using these channel gains, transmit power levels that each UE should allocate to the two network interfaces and the corresponding throughputs are determined. In the third time-scale, resources of the cellular network are reallocated based on these power and throughput levels, using instantaneous channel gains.

Resources are allocated in two steps in order to simplify allocation of non-orthogonal resources and calculation of transmission power levels for D2D and traditional links which share the non-orthogonal resources. In the first step, traditional links and the D2D links which use orthogonal resources are allocated resources. In the second step, remaining D2D links are allocated non-orthogonal resources. $\mathcal{S}_1$ and $\mathcal{S}_2$ denote sets of the users who are allocated resources during the first and the second steps respectively, where $\mathcal{S}_1 \cap \mathcal{S}_2 = \mathcal{S}_N$ and $\mathcal{S}_1 \cup \mathcal{S}_2 = \emptyset$.

The D2D links in WLANs are allocated orthogonal resources as reusing resources for D2D and traditional links in proximity is inefficient. Similarly, in the cellular network, D2D links within a $d_L$ distance from the eNB are allocated orthogonal resources, as these D2D links cause severe CCI to the eNB if they are allocated non-orthogonal uplink resources. In the cellular network, D2D links which are at least a $d_L$ distance away from the eNB and are belongs to multi-homing users, are also allocated orthogonal resources, because they should be allocated resources during the first step due to following reason. Equation (2.5) shows that the transmit power levels of all the users in a WLAN are correlated, because an user's throughput via contention-based channel access depends on the transmit power levels of all the users in the WLAN. Also, these users could be currently multi-homing. That is, they could also be accessing the cellular network using a part of their transmit power. Therefore, all the multi-homing users should be jointly allocated resources with other WLAN users, during the first step. Remaining D2D links in the cellular network are allocated non-orthogonal resources realizing reuse gain in the interworking system. Second time-scale joint resource allocation algorithm is shown in Fig. 4.5.

### 4.3.2.1 First Step of the Second Time-Scale Joint Resource Allocation

Resources of the cellular network and the WLANs are jointly allocated in the first step, subject to C9–C11 while assuming eNB receives the worst CCI of $I_c$ from D2D links. This step is performed via *Steps 3–6* in Fig. 4.5, where $R_i^C$ and $R_i^W$ denote throughputs achieved by the $i$th user ($i \in \mathcal{S}_1$) through the cellular network and a WLAN using the modes which have already been selected by the mode selection algorithm. For example, if D2D mode has been selected for the $i$th user to communicate over the cellular network, then $R_i^C = R_i^{C(D)}$. To allocate resources in this step, Algorithm 3 derived in Chap. 3 is used with three modifications: (1) to further reduce time complexity of Algorithm 3, $\mathcal{S}^{CB}$ is assumed to be all the users in the WLAN; (2) instead of using average channel gains of the channels over the WLAN, instantaneous channel gains are used; and (3) only one QoS constraint for each user is considered, and it corresponds to the data traffic QoS constraint with the dual variable $\lambda$ in Algorithm 3.

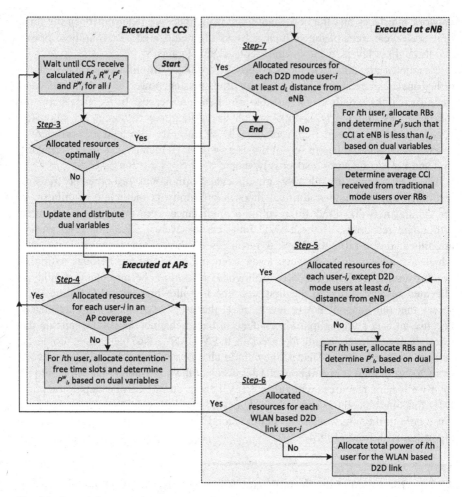

**Fig. 4.5** Second time-scale joint resource allocation algorithm

In *Step 3* in Fig. 4.5, dual variables $\boldsymbol{\lambda}$ and $\boldsymbol{\mu}$, which correspond to QoS and total power constraints of the problem $\mathcal{P}6$ respectively, are updated using the subgradient method [10, 51, 52]. To update the dual variables, eNB and each AP feedback $R_i^C$, $R_i^W$, $P_i^C$ and $P_i^W$ calculated using the current values of the dual variables. Moreover, when $R_i^C$ and $P_i^C$ are calculated for the links which use traditional mode, additional interference of $I_c$ is taken into account.

In *Step 4*, first the APs determine $\bar{P}_{i,j}^{CF}$, $\rho_{i,j}^{CF}$ and $P_i^{CB}$ for $i \in \mathcal{S}_1$ by (3.18)–(3.21), (3.23), and (3.24). Second, APs calculate

$$R_i^W = \frac{T_{CF}}{T_P} \sum_{j \in \mathcal{K}^{CF}} \rho_{i,j}^{CF} R_{i,j}^{CF}(\bar{P}_{i,j}^{CF}) \qquad (4.1)$$

and

$$P_i^W = P_{avg,i}^{CB}(\mathbf{P}^{CB}) + \frac{T_{CF}}{T_P} \sum_{j \in \mathcal{K}^{CF}} \bar{P}_{i,j}^{CF}, \tag{4.2}$$

where $R_{i,j}^{CF}(\bar{P}_{i,j}^{CF})$ and $P_{avg,i}^{CB}(\mathbf{P}^{CB})$ are given by (2.2) and (2.7).

In *Step 5*, first the eNB determines $\rho_{i,k}^C$ and $\bar{P}_{i,k}^C$ using (3.27) and (3.28). Second, it calculates

$$R_i^C = \sum_{k \in \mathcal{K}^C} \rho_{i,k}^C R_{i,k}^C(\bar{P}_{i,k}^C) \tag{4.3}$$

and

$$P_i^C = \sum_{k \in \mathcal{K}^C} \bar{P}_{i,k}^C, \tag{4.4}$$

where $R_{i,k}^C(\bar{P}_{i,k}^C)$ is given by (2.1).

In *Step 6*, resources for the WLAN based D2D links, which are setup by sending the control signals through the cellular network, are allocated. For these links, $P_i^W = P_{T,i}$ and $R_i^W$ is given by (2.2), as explained in Sect. 4.3.1.

#### 4.3.2.2  Second Step of the Second Time-Scale Joint Resource Allocation

The second step is performed via *Step 7* in Fig. 4.5. In this step, D2D links of the users in $\mathcal{S}_2$ are allocated cellular network resources (i.e., subcarriers) and transmit power subject to QoS and total power constraints. The subcarriers allocated in this step have already been allocated to traditional mode links during the first step of the second time-scale joint resource allocation algorithm. Therefore, when subcarriers are allocated to the D2D links in this step, CCI received over the subcarriers is calculated and taken into account. Furthermore, the transmit power levels of the D2D link transmitters are determined such that they do not exceed CCI of $I_c$ at the eNB. To reduce the required number of channel estimations to calculate received and caused CCI, average channel gains which can be estimated based on distances are used. Distances between the users and the eNB can be calculated using two techniques that are supported by the LTE cellular networks: (1) assisted global navigation satellite systems (A-GNSS) positioning, and (2) observed time difference of arrival (OTDOA) positioning. Position information can be exchanged between the users and the eNB via LTE positioning protocol (LPP).

In *Step 7* in Fig. 4.5, resources are allocated using the algorithm used for allocating resources in *Steps 3* and *5*. In this step, the dual variables $\lambda$ and $\mu$ are updated only considering $R_i^C$ and $P_i^C$, as these users communicate only via the cellular network. First, from (3.27), the eNB calculates [66]

$$P_{i,k}^C = \min \left\{ \frac{I_c}{(h_{i,k}^C)^2}, \left[ \frac{\Delta f}{\ln(2)} \frac{(1 + \lambda_i)}{\mu_i} - \frac{1}{\alpha_{i,k}^C} \right]^+ \right\}, \ \forall i \in \mathcal{S}_2, k \in \mathcal{K}^C, \quad (4.5)$$

where $h_{i,k}^C$ is the average channel gain of the channel between the $i$th user and the eNB over the $k$th subcarrier, and $\alpha_{i,k}^C$ is calculated taking the received CCI into account. Second, $\rho_{i,k}^C, \forall i \in \mathcal{S}_2, k \in \mathcal{K}^C$ are calculated by substituting $P_{i,k}^C$ into (3.28). Third, $R_i^C$ and $P_i^C$ are determined from (4.3) and (4.4), where $\bar{P}_{i,k}^C = \rho_{i,k}^C P_{i,k}^C$.

### 4.3.3  Third Time-Scale: Cellular Network Resource Allocation

Resources of the cellular network are reallocated in the third time-scale using instantaneous channel gains. Reallocation of resources in this fast time-scale also allows capturing multiuser diversity over the fast fading wireless channels in the cellular network. In this time-scale, resources are allocated using the same two-step approach used in the second time-scale. In the first step, the $i$th multi-homing user has total power of $P_i^C$ to communicate over the cellular network, and requires minimum rate of $R_{min} - R_i^W$ via the cellular network, where $P_i^C$ and $R_i^W$ were calculated in the second time-scale, and $R_{min} = R_{min,i}^{ND}$ if the $i$th user requires non-D2D communication, or $R_{min} = R_{min,i}^{D2D}$ otherwise. Second step remains unchanged.

## 4.4  Implementation of the Resource Allocation Scheme

This section describes a semi-distributed implementation of the proposed resource allocation scheme in an interworking system which consists of an LTE-A network and IEEE 802.11n WLANs operating in 2.1 and 2.4 GHz frequency bands, respectively. Implementation of the resource allocation scheme is shown in Fig. 4.6. APs, eNB and CCS perform different functions of the resource allocation scheme. CCS is connected to the LTE-A EPC through the packet data network gateway (PDN-GW). It communicates with the eNB through serving gateway (S-GW) and PDN-GW, and with the APs through WLAN access gateway (WAG), evolved packet data gateway (ePDG) and PDN-GW. The key advantages of this implementation are: (1) signaling overhead and signaling delay are reduced; (2) computational burden is distributed over the networks; and (3) a single point of failure is eliminated.

The CCS performs mode selection. It receives average channel gains of the traditional and potential D2D links over both networks to select the modes. Once it selects the modes, it sends information of the selected modes to the APs and the eNB to setup links.

The first step of the second time-scale resource allocation jointly allocates cellular network and WLAN resources. In this step, the eNB allocates the cellular network resource while the APs allocate each WLAN's resources. CCS controls

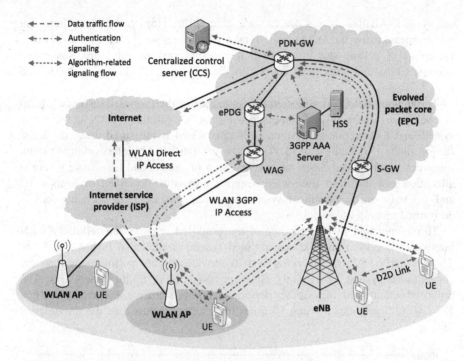

**Fig. 4.6** Semi-distributed implementation of the resource allocation scheme

how the eNB and the APs allocate resources such that resource allocation for the entire interworking system iteratively converges to the global optimum. Specifically, CCS broadcasts the dual variables $\lambda$ and $\mu$ to the eNB and the APs. Next, the eNB and the APs eNB allocate resources based on the received dual variables, and send $P_i^C$, $P_i^W$, $R_i^C$ and $R_i^W$, $\forall i \in S_1$ to the CCS. Finally, CCS updates the dual variables and redistributes to the eNB and the APs. As shown in *Step 3* of Fig. 4.5, this process continues until the resource allocation reaches the global optimum.

The eNB performs the second step of the second time-scale resource allocation and the third time-scale resource allocation, as only the cellular network resources are allocated in these resource allocations. Furthermore, implementing these two resource allocations entirely in the eNB provides a low signaling delay as information is not required to be exchanged with the CCS. A low signaling delay is essential for the third time-scale operations due to very short time slot duration.

## 4.5  Performance Evaluation

Performance of the proposed resource allocation scheme is evaluated in the system described in Sect. 3.8 with following modifications. Twenty-five high mobility and twenty-five low mobility users are uniformly and randomly distributed in the system. All these users are capable of multi-homing, and $Y\%$ of them can communicate using the D2D mode. Total power available at each user is 27 dBm. Durations of a time slot in the first, the second and the third time-scales are 640 ms, 64 ms and 1 ms, respectively. In the LTE-A network, $d_L = 200$ m and $I_c = -62$ dBm.

Throughput and QoS satisfaction performance of the proposed resource allocation scheme is evaluated and compared with that of a cellular/WLAN interworking system and a conventional system. QoS satisfaction is measured using the satisfaction index (SI) defined in (3.37). In the cellular/WLAN interworking system, resources are allocated based on the first step of the second time-scale resource allocation and the third time-scale resource allocation described in Sects. 4.3.2 and 4.3.3. In the conventional system, resource allocation for each network is performed separately.

Throughput and SI performance of the proposed scheme, the cellular/WLAN interworking system and the conventional system is shown in Figs. 4.7 and 4.8. The cellular/WLAN interworking system provides higher performance than the conventional system as it jointly allocates resources of multiple networks. The proposed scheme provides further performance improvements. For example, when $Y = 40\%$, it provides 3.4 and 10 times higher throughputs than the throughputs

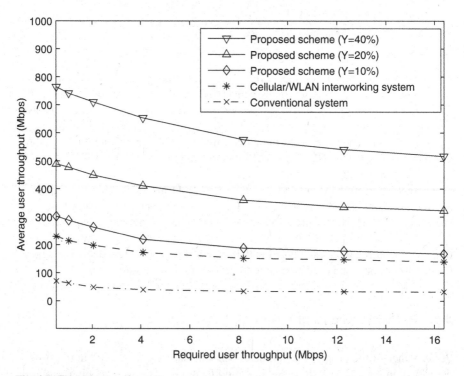

**Fig. 4.7** Throughput performance

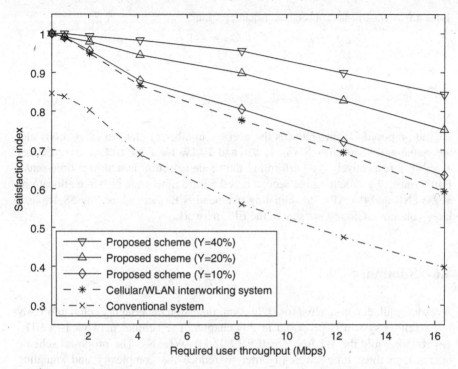

**Fig. 4.8** QoS satisfaction

achieved in the cellular/WLAN interworking system and the conventional system, respectively. It provides such enhanced performance due to five reasons: (1) joint allocation of resources in multiple networks; (2) exploitation of strong wireless channels available between the users in proximity; (3) realization of hop and reuse gains; (4) setup WLAN based D2D links with assistance from the cellular network; and (5) efficient use of orthogonal and non-orthogonal resources to manage interference. Furthermore, the throughput and QoS performance of the proposed scheme increases with $Y$, as more D2D links can be established when $Y$ increases.

Complexity of the proposed resource allocation scheme is measured in terms of the number of iterations required per user for the first step of the second time-scale resource allocation. Only the first step of the second time-scale resource allocation is considered for measuring the complexity, as it has the highest complexity due to jointly allocating cellular network and WLAN resources. Average number of iterations required for the proposed scheme is shown in Table 4.1. Number of iterations required per user decreases with $Y$, as more users are allocated WLAN based D2D links and that more D2D links are allocated non-orthogonal cellular network resources during the second step. However, it increases with $R_{min}$, as more iterations are required to find the dual variables which ensure satisfaction of this high minimum data rate requirement.

**Table 4.1** Average number of iterations required per user

|            | $R_{min} = 512$ kbps | $R_{min} = 4$ Mbps | $R_{min} = 16$ Mbps |
|------------|----------------------|--------------------|---------------------|
| $Y = 10\%$ | 5.45                 | 6.51               | 7.93                |
| $Y = 20\%$ | 5.23                 | 6.16               | 7.62                |
| $Y = 40\%$ | 4.80                 | 5.64               | 6.91                |

The proposed scheme reduces the average number of channel estimations and the signaling overhead by 8.3%, 15.9% and 29.1% for $Y = 10\%$, $Y = 20\%$ and $Y = 40\%$ respectively, by performing the mode selection in a slower time-scale. Furthermore, by executing the second and the third time-scale resource allocations at the eNB and the APs, the signaling overhead is further reduced by 58.4% as a large volume of CSI are not sent to the EPC network.

## 4.6   Summary

A resource allocation scheme for D2D communication underlaying cellular/WLAN interworking system is proposed in this chapter. The cellular network is a LTE-A network while the WLANs are IEEE 802.11n WLANs. The proposed scheme operates on three time-scales in order to reduce the complexity and signaling overhead caused by the mode selection process and to allocate cellular network and WLAN resources based on the underlying MAC and PHY technologies. The proposed scheme is implemented in a semi-distributed manner to further reduce the signaling overhead and delay while preventing a single point of failure. Simulation results demonstrate that the proposed scheme significantly improves the system throughput and the QoS satisfaction, by interworking of multiple networks and enabling D2D communication.

# Chapter 5
# Resource Allocation for Interworking Macro Cell and Hyper-Dense Small Cell Networks

This chapter investigates uplink resource allocation for interworking macro cell and hyper-dense small cell networks. Interworking among these two networks is considered as it provides several advantages: (1) it increases network throughput; (2) it enhances network coverage by merging coverages of the two networks; and (3) it provides better QoS by reducing call drops of highly mobile users, which is achieved by allocating users to the networks based on the user mobility levels. The main challenge for allocating resources for this system is the high complexity of resource allocation algorithms due to resources of a large number of cells need to be jointly allocated considering co-channel interference (CCI) among the small cells, time-varying network loads and limited capacity backhauls. As computational capacity is limited in small cell BSs, use of cloud computing to execute the resource allocation algorithms is also investigated in this chapter.

This chapter first discusses the challenges for resource allocation in this system and the related work in literature. Second, it presents the macro cell and hyper-dense small cell interworking system model. Third, a resource allocation algorithm that operates on two time-scales using cloud computing is derived. Fourth, performance of the derived resource allocation algorithm is evaluated via simulations.

## 5.1 Challenges for Resource Allocation

The main challenge for allocating resources in this interworking system is the high complexity of the resource allocation algorithms. High complexity of these algorithms is a result of jointly allocating a large amount of resources subject to a large number of constraints, such as CCI and backhaul capacity constraints. In uplink communication, these resources consist of frequency resources of small cells and macro cells, and transmit power resources of the users. As there are a large number of small cells in a hyper-dense small cell network, there are a large

© The Author(s) 2018
A.T. Gamage, X.(S.) Shen, *Resource Management for Heterogeneous Wireless Networks*, SpringerBriefs in Electrical and Computer Engineering,
DOI 10.1007/978-3-319-64268-0_5

number of resources in the interworking system. There are a large number of CCI constraints as there is a large number of small cells which reuse a small set of frequencies. Moreover, considering CCI during resource allocation is a necessity as most of these small cells are deployed in an unplanned manner by the subscribers. Consideration of backhaul capacities during resource allocation is important in order not the bottleneck the limited capacity backhauls of the small cell BSs.

These complex resource allocation algorithms can be executed using vastly available cloud computing resources, as the limited computational capacity in small cell BSs is not sufficient. To execute the resource allocation algorithms, a dedicated centralized server at the mobile core network can also be used. However, use of cloud computing provides several advantages compared to using a dedicated centralized server [67, 68]: (1) ability to adapt to varying computational requirements; (2) low operational and maintenance costs; and (3) high reliability with spatially distributed multiple redundant servers.

A key challenge to use cloud computing to execute a resource allocation algorithm is the delay when access the cloud computing facility [69]. This delay is caused by transmission delay, queuing delay at the cloud and processing time at cloud computing servers. Typically the total delay is in the order of 100s of milliseconds, and it increases with the size of the computing task. However, resource allocation decisions need to be made within a few milliseconds as they should be adapted to rapidly varying wireless channel conditions.

## 5.2   Related Work

Resources of this interworking system should be allocated such that CCI in the system stays within a tolerable level. To manage CCI, three types of techniques are proposed in literature: (1) interference avoidance techniques; (2) diversity combining and interference suppression techniques; and (3) interference controlling techniques.

Interference avoidance schemes include resource partitioning schemes, such as Almost Blank Subframes (ABS) allocation and Fractional Frequency Reuse (FFR) techniques. FFR techniques eliminate inter cell CCI by allowing users at cell edges to utilize only a sub-set of available frequency channels, while users at cell centers utilize all the frequency channels [70, 71]. The sub-sets of frequency channels for neighboring cell edge users are determined such that they do not overlap. This technique is less efficient in hyper-dense small cell networks, as unplanned cell deployments and small cell sizes prevent using all the frequency channels in cell centers and require partitioning the available frequency channels into a large number of non-overlapping sub-sets. Consequently, frequency reuse is significantly reduced. ABS allocation techniques schedule transmissions within small cells during ABS transmissions of the macro cell to avoid CCI [72, 73]. Though ABS allocation eliminates CCI between the macro and small cells, it does not eliminate inter cell CCI among the small cells.

Diversity combining and interference suppression techniques for uplink include joint decoding and network multiple-input-and-multiple-output (MIMO) techniques. Joint decoding techniques improve SINR due to diversity gain. That is, they decode users' data combining signals received by several BSs using techniques such as selection diversity and maximal ratio combining (MRC). Network MIMO techniques achieve significantly higher performance compared to diversity combining techniques [74], and performance of network MIMO techniques is investigated in [75]. Network MIMO techniques based on zero forcing and minimum mean squared error (MMSE) equalizers are proposed in [28, 76, 77]. To reduce the amount of control information transmitted to UEs, transmission of precoding matrix indexes is proposed in [78]. These diversity combining and interference suppression techniques cannot be used in hyper-dense small cell networks due to three reasons. First, dense unplanned small cell deployments cause each cell to receive CCI not only from neighboring cells, but also from neighbors of the neighboring cells. Second, limited capacity small cell backhauls cannot be used for transmission of instantaneous channel state information (CSI) of a large number of users. For each BS, there are a large number of users, who are associated with adjacent small cells, in vicinity. The instantaneous CSI of these users is required to process these users' signals received by different BSs [73, 79]. Third, high cost of antenna arrays.

Several interference controlling schemes based on allocating resources, such as transmit power, subcarriers and time slots, are proposed in [79–81]. In [79], employing small cell BSs as relays for uplink macro cell communications to reduce CCI is investigated. In this scheme, transmit power levels and amount of data routed via relays are determined by solving a non-cooperative game among users. Power and subcarrier allocation to reduce CCI caused by femto cell users to macro cell users is investigated in [80]. A distributed algorithm to optimize power and frequency resource allocation subject to inter cell CCI constraints is proposed in [81]. Since resource allocation algorithms for interworking macro cell and hyper-dense small cell networks system are intended to be executed using cloud computing, further investigation on resource allocation schemes to take into account the high delay when access cloud computing resources is required.

To overcome the effect of high delay when access cloud computing resources, predicting future wireless channel states using a finite state Markov chain (FSMC) is proposed in [82]. At each resource allocation interval, network resources are allocated using cloud computing and based on predicted channel states. Channel states are predicted using CSI received a few resource allocation intervals ago. Such scheme cannot be directly implemented in the interworking macro cell and hyper-dense small cell networks system due to three reasons. First, number of channel states in the FSMC is very large due to existence of a large number of BSs and users in the hyper-dense small cell network. Second, transmission of CSI during each resource allocation interval, which is about 1 ms for these cellular networks, consumes a large portion of available wireless bandwidth and backhaul capacity. Third, determining resource allocation decisions for this interworking system takes a longer time duration than the resource allocation interval of about 1 ms. Due to the second and the third reasons, this interworking system requires a resource allocation scheme that allocates resources once for a few resource allocation intervals, and yet achieves high performance. Thus, in this chapter a resource allocation scheme that overcomes these drawbacks by operating on two time-scales and using time-correlated wireless channels is derived.

## 5.3   Macro Cell and Hyper-Dense Small Cell Interworking System Model

System model focuses on uplink resource allocation in first and third type areas described in Sect. 2.1. In these areas, macro and small cells are deployed in planned and unplanned manners, respectively. Interworking system that consists of these two networks is divided into $C$ clusters. To reduce complexity of resource allocation, resources of different clusters are allocated separately, while resources of a cluster are jointly allocated. A cluster is shown in Fig. 5.1. The $c$th cluster consists of $\mathcal{M}^{(c)}$ set of macro cells and $\mathcal{N}^{(c)}$ set of small cells. $\mathcal{U}^{(c)}$ denotes the set of users in the $c$th cluster. Each user connects to the cell (i.e., BS) that is closest to the user. Set of users connects to the $b$th cell is denoted by $\mathcal{U}_b^{(c)}$, where $b \in \mathcal{M}^{(c)} \cup \mathcal{N}^{(c)}$. Total available frequency band is divided into subcarriers with bandwidth of $\Delta f$, and $\mathcal{K}$ denotes the set of subcarriers. Set of subcarriers used by a cell is dynamically determined based on cell load and CCI among the cells.

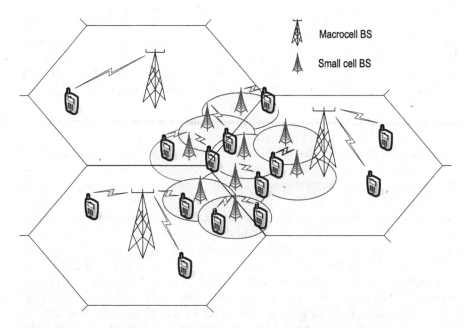

**Fig. 5.1**  A cluster of the interworking macro cell and hyper-dense small cell networks

### 5.3.1 Cloud Assisted Two-Time-Scale Resource Allocation

To maximize aggregated throughput of users in a cluster, resource allocation decisions are made at cloud (i.e., at cloud computing servers) as well as at BSs. First, user allocation, subcarrier allocation and water-levels are determined at the cloud. To make these decisions at the cloud, the BSs send information about channels between the BSs and the users to the cloud. Second, the cloud distributes these decisions to the BSs. Third, based on the water-levels and instantaneous CSI, the BSs determine user transmit power levels. As shown in Fig. 5.2, the cloud and the BSs make decisions in a slow and a fast time-scale, respectively.

Time slot durations of the fast and the slow time-scales are denoted by $T_F$ and $T_S$, respectively. $T_S > T_F$, and there are $L$ fast time-scale time slots within each slow time-scale time slot. $T_F$ is determined based on the resource allocation interval of cellular networks (see Sect. 3.1). $T_S$ is determined based on two factors: (1) total delay when access the cloud, which depends on the cluster size; and (2) acceptable control signaling overhead, as CSI and resource allocation decisions are transmitted from BSs to the cloud and from the cloud to the BSs at every slow time-scale time slot. $T_0$ denotes the difference between $T_S$ and the time duration requires to transmit CSI to the cloud, determine resource allocation decisions at the cloud and send the decisions back to the BSs. For simplicity, $L$ and $L_0$ are assumed to be integers, where $L = T_S/T_F$ and $L_0 = T_0/T_F$.

The decisions made at the cloud during current slow time-scale time slot are used in the BSs during next slow time-scale time slot. In the slow time-scale, first the users are allocated to the cells. Next, the subcarrier allocation and the water-levels for the allocated subcarriers are determined by solving a joint resource allocation problem, based on average CCI, average user throughputs and average user transmit power estimated over the next slow time-scale time slot. The averages of CCI, user throughputs and user transmit power are estimated using channel statistics and instantaneous CSI at the $L_0$th fast time-scale time slot within the current slow time-scale time slot, and by taking the fast time-scale resource allocation into account. In the fast time-scale, transmit power levels are determined based on the instantaneous CSI and the already determined water-levels, using a water filling algorithm.

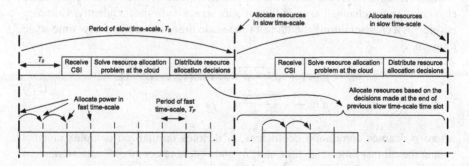

**Fig. 5.2** Two time-scale cloud assisted resource allocation

### 5.3.2  Cloud Access Model

The cloud computing resources (i.e., servers) are shared by several users and services to perform different computing tasks. Thus, total delay ($d_{Total}$) when access the cloud is the sum of following three components: (1) transmission delay ($d_t$); (2) queuing delay ($d_q$); and (3) processing time ($d_p$). $d_t$ for the $c$th cluster is the maximum round-trip-time between a BS in the cluster and the cloud servers, and is given by [83]

$$d_t = \max_{\forall b \in \mathcal{M}^{(c)} \cup \mathcal{N}^{(c)}} \{2 \times 10^{-8} l_b + 5 \times 10^{-3}\}, \tag{5.1}$$

where $l_b$ denotes the distance between the $b$th BS ($b \in \mathcal{M}^{(c)} \cup \mathcal{N}^{(c)}$) and cloud servers.

When computing tasks are sent to the cloud, they are queued until required amount of computing resources are freed up. Waiting time in this queue is referred to as the queuing delay. When the cloud utilization is low, the queue can be approximated by a M/G/1 queue. Then, by assuming an exponential processing (i.e., service) time distribution [84],

$$d_q = 1/\varphi_s + \frac{(\chi_s^2 + 1/\varphi_s^2)\xi}{2(1 - \tau)}, \tag{5.2}$$

where $1/\varphi_s$ and $\chi_s^2$ are the mean and variance of the processing time distribution, respectively; $\xi$ is the average arrival rate of computing tasks; and $\tau = \xi/\varphi_s$. Processing time is the amount of time taken by the cloud servers to complete a computing task. It is given by $\alpha_s + v_s$, where $\alpha_s$ is a non-negative constant and $v_s$ is an exponentially distributed random variable as explained in [84].

### 5.3.3  Channel Model

Wireless channels over subcarriers are modeled as time-correlated Rayleigh fading channels. These channels over different subcarriers fade independently. Complex envelop of a channel during the $t$th fast time-scale time slot within a slow time-scale time slot is

$$\begin{aligned} \tilde{h}_t &= \rho \tilde{h}_{t-1} + \sqrt{1 - \rho^2} \tilde{w}_t \,, \ t \in \{1, \ldots, L - 1\} \\ &= \rho^t \tilde{h}_0 + \sqrt{1 - \rho^{2t}} \tilde{w}, \end{aligned} \tag{5.3}$$

where $\rho$ denotes correlation coefficient; $\tilde{h}_t$ denotes normalized complex channel gain at the $t$th fast time-scale time slot; $\tilde{w}_t$ and $\tilde{w}$ denote complex Gaussian random variables, i.e., $\tilde{w}_t, \tilde{w} \sim \mathcal{CN}(0, \sigma^2)$; and $\sigma^2$ denotes the ratio of average power gain of the channel to the additive white Gaussian noise power. The normalized complex channel gain is the ratio of the channel gain to the square root of the noise power. The second line of (5.3) is obtained due to the fact that a sum of independent Gaussian random variables is also a Gaussian random variable. Next, $\tilde{h}_t$ can be rewritten as

$$\tilde{h}_t = (\rho^t h_{I,0} + \sqrt{1 - \rho^{2t}} w_I) + j(\rho^t h_{Q,0} + \sqrt{1 - \rho^{2t}} w_Q), \tag{5.4}$$

where $h_{I,0}$ and $w_I$ are in-phase components of $\tilde{h}_0$ and $\tilde{w}$, respectively; and $h_{Q,0}$ and $w_Q$ are quadrature-phase components of $\tilde{h}_0$ and $\tilde{w}$, respectively. Envelop of the channel gain is then given by

$$|\tilde{h}_t|^2 = X_I^2 + X_Q^2, \tag{5.5}$$

where $X_I = \rho^t h_{I,0} + \sqrt{1 - \rho^{2t}} w_I$ and $X_Q = \rho^t h_{Q,0} + \sqrt{1 - \rho^{2t}} w_Q$. Furthermore, $X_I \sim \mathcal{N}(\rho^t h_{I,0}, (1 - \rho^{2t})\sigma^2/2)$ and $X_Q \sim \mathcal{N}(\rho^t h_{Q,0}, (1 - \rho^{2t})\sigma^2/2)$. Since the variances of $X_I$ and $X_Q$ are identical, the conditional probability $Pr(|\tilde{h}_t|^2||\tilde{h}_0|^2)$ is a noncentral Chi-square distribution with two degrees of freedom [85]. Thus,

$$Pr(|\tilde{h}_t|^2 = y||\tilde{h}_0|^2) = \frac{1}{2\bar{\sigma}^2} e^{-(s^2+y)/2\bar{\sigma}^2} \sum_{k=0}^{\infty} \frac{y^k (s/2\bar{\sigma}^2)^{2k}}{(k!)^2}, \tag{5.6}$$

where $\bar{\sigma}^2 = (1 - \rho^{2t})\sigma^2/2$ and $s = \sqrt{(\rho^t h_{Q,0})^2 + (\rho^t h_{Q,0})^2} = \rho^t |\tilde{h}_0|$. Based on (5.6) and information of the channels, averages of CCI, users' throughputs and users' transmit power over the current or a future slow time-scale time slot can be determined.

## 5.4 Fast Time-Scale Resource Allocation

In the fast time-scale, user transmit power levels are calculated based on instantaneous CSI using a water filling algorithm which is different from the traditional water filling algorithm. In the traditional algorithm, there is a fixed water-level for all the subcarriers, whereas in the algorithm used in this chapter has a different water-level for each subcarrier and these water-levels remain fixed over a slow time-scale time slot. An example of the water-levels of this algorithm is shown in Fig. 5.3. Whenever a water-level is higher than ratio of noise plus interference power to channel power gain, transmit power is allocated for that subcarrier. Water-levels of different subcarriers are different as the transmit power levels are calculated considering CCI introduced to and received from other cells.

From (2.1), throughput of the $u$th user during the $t$th fast time-scale time slot over the $k$th subcarrier is

$$r_{ukt} = \Delta f \log_2 \left(1 + \frac{H_{ukt} p_{ukt}}{1 + I_{ukt}}\right), \tag{5.7}$$

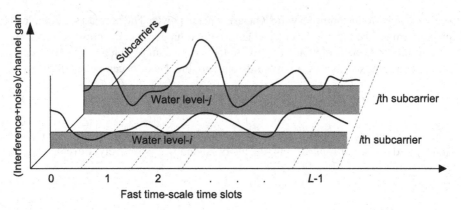

**Fig. 5.3** Water-levels of the water filling algorithm

where $H_{ukt}$ is the ratio of channel power gain (i.e., $|\tilde{h}_t|^2$) to noise level; $p_{ukt}$ is the transmit power level; and $I_{ukt}$ is the ratio of interference (i.e., CCI) power to noise power. $H_{ukt}$, $p_{ukt}$ and $I_{ukt}$ are for the $u$th user over the $k$th subcarrier during the $t$th fast time-scale time slot. To simplify the resource allocation, interference level (i.e., $I_{ukt}$) received by an user over an slow time-scale time slot is assumed to be the same, and equivalent to the average interference ($I_{uk}$) determined at the cloud. Then, from the water filling algorithm, the optimal transmit power levels are

$$p^{*}_{ukt} = \left[ \mu_{uk} - \frac{1 + I_{uk}}{H_{ukt}} \right]^{+}, \qquad (5.8)$$

where $\mu_{uk}$ is the water-level for the $u$th user over the $k$th subcarrier. To determine $p^{*}_{ukt}$ at the BSs, the cloud sends $\mu_{uk}$ and $I_{uk}$ to the BSs.

## 5.5   Slow Time-Scale Resource Allocation

Resource allocation decisions (i.e., user allocation, subcarrier allocation and water-levels) made during the current slow time-scale time slot are used in the network during the next slow time-scale time slot. To determine these decisions, averages of user throughputs, power consumptions and CCI over the next slow time-scale time slot should be estimated. These averages for the $u$th user over the $k$th subcarrier are denoted by $R_{uk}$, $P_{uk}$ and $I_{uk}$, respectively. Furthermore, $R_{uk}$, $P_{uk}$ and $I_{uk}$ should be estimated considering the fast time-scale power allocation at the BSs, as variations in transmit power levels affect them.

Derivations of $P_{uk}$, $R_{uk}$ and $I_{uk}$ based on time-correlated Rayleigh fading channels, which statistically model the relationship between channel gains at different time instances, are given in Appendices B.1, B.2 and B.3, respectively. However, solving slow time-scale resource allocation problem using expressions in these appendices is highly complex due to two reasons. First, $R_{uk}$ is a non-convex function of the water-levels. Thus, the resource allocation problem becomes non-convex and requires exhaustive search methods to solve the problem. Second, since expressions for $P_{uk}$, $R_{uk}$ and $I_{uk}$ are highly complex, it is not computationally feasible to calculate $P_{uk}$, $R_{uk}$ and $I_{uk}$ for a large number of search points in search algorithms. Thus, to gain valuable insights on the effectiveness of the cloud assisted two-time-scale resource allocation, in this chapter the slow time-scale resource allocation problem is solved with two assumptions: (1) $P_{uk}$, $R_{uk}$ and $I_{uk}$ are calculated assuming the channels are not time-correlated, i.e., $\rho = 0$; and (2) the transmit power levels are assumed to remain fixed within a slow time-scale time slot.

### 5.5.1 User Allocation

The first step of the slow time-scale resource allocation is to associate users with BSs. Each user is allocated to the BS to which the user has the strongest channel gain. Such user allocation policy is optimal as the resource allocation scheme derived in this chapter dynamically adjusts the bandwidth available at each BS based on the BS's load. Dynamically allocating bandwidths to BSs is more efficient than statically allocating bandwidths specially when network load significantly changes and moves across the network with time. A drawback of this user allocation is that it may saturate some BSs' backhauls.

### 5.5.2 Subcarrier Allocation and Water-Level Calculation

Subcarrier allocation and water-levels are determined assuming channels are not time-correlated and transmit power levels remain unchanged within a slow time-scale time slot. Therefore, from (5.6), (B.10) and (B.11), average throughput of the $u$th user ($u \in \mathcal{U}^{(c)}$) is

$$
\begin{aligned}
R_{uk} &= \frac{1}{L} \sum_{t=0}^{L-1} \int_0^\infty \Delta f \log_2 \left( 1 + \frac{H_{ukt} P_{uk}}{1 + I_{uk}} \right) \frac{1}{\sigma_{uk}^2} \exp \left( \frac{-H_{ukt}}{\sigma_{uk}^2} \right) dH_{ukt} \\
&= \frac{1}{L} \sum_{t=0}^{L-1} \frac{\Delta f}{\ln(2)} \exp \left( \frac{1 + I_{uk}}{\sigma_{uk}^2 P_{uk}} \right) E_1 \left( \frac{1 + I_{uk}}{\sigma_{uk}^2 P_{uk}} \right),
\end{aligned}
\tag{5.9}
$$

where $\sigma_{uk}^2$ is the average normalized power gain of the channel between the $u$th user and the BS to which the $u$th user is connected to, over the $k$th subcarrier; and $E_1(\theta)$ is the exponential integral given by (3.30). Normalized power gain is the ratio of channel power gain to noise power. A tight lower bound for $E_1(\theta)$ is given by [86]

$$
E_1(\theta) > 0.5 \exp(-\theta) \ln(1 + 2/\theta). \tag{5.10}
$$

Therefore,

$$R_{uk} \approx \frac{1}{L} \sum_{t=0}^{L-1} \frac{\Delta f}{2 \ln(2)} \ln\left(1 + \frac{2\sigma_{uk}^2 P_{uk}}{1 + I_{uk}}\right)$$

$$= \frac{\Delta f}{2} \log_2\left(1 + \frac{2\sigma_{uk}^2 P_{uk}}{1 + I_{uk}}\right). \tag{5.11}$$

Next, from (B.16) and (B.17),

$$I_{uk} = \frac{1}{L} \sum_{t=0}^{L-1} \sum_{v \in \mathcal{U}^{(c)} \backslash u} \int_0^\infty H_{vkt}^{(u)} P_{vk} \frac{1}{(\sigma_{vk}^{(u)})^2} \exp\left(\frac{-H_{vkt}^{(u)}}{(\sigma_{vk}^{(u)})^2}\right) \mathrm{d}H_{vkt}^{(u)}$$

$$= \sum_{v \in \mathcal{U}^{(c)} \backslash u} (\sigma_{vk}^{(u)})^2 P_{vk}, \tag{5.12}$$

where $H_{vkt}^{(u)}$ is the normalized power gain of the channel between the $v$th user and the BS to which the $u$th user is connected to, over the $k$th subcarrier during the $t$th fast time-scale time slot within the next slow time-scale time slot; and $(\sigma_{vk}^{(u)})^2$ is the average normalized power gain of the same channel.

The resource allocation problem to determine subcarrier allocation and water-levels for the $c$th cluster such that aggregated users' throughput is maximized is

$$\mathcal{P}7 : \max_{P} \qquad \sum_{u \in \mathcal{U}} \sum_{k \in \mathcal{K}} R_{uk}$$

$$\text{s.t.} \qquad \text{C12} : (5.11)$$

$$\text{C13} : (5.12)$$

$$\text{C14} : \sum_{k \in \mathcal{K}} P_{uk} \le P_{T,u}, \ \forall u \in \mathcal{U}^{(c)}$$

$$\text{C15} : P_{uk} \ge 0, \ \forall u \in \mathcal{U}^{(c)}, k \in \mathcal{K},$$

where $P$ is a vector consisting of all the transmit power variables and $P_{T,u}$ is the $u$th user's total available power. When $P_{uk}^* > 0$, the $k$ subcarrier is allocated to the $u$th user. If $P_{vk}^* > 0, \forall v \in \mathcal{V}$, where $\mathcal{V} \subseteq \mathcal{U}^{(c)}$, then all the users in $\mathcal{V}$ simultaneously transmit over the $k$ subcarrier. From (5.11) and (5.12), the water-levels are given by

$$\mu_{uk} = P_{uk}^* + \frac{1 + I_{uk}}{\sigma_{uk}^2}, \ \forall u \in \mathcal{U}^{(c)}, k \in \mathcal{K}. \tag{5.13}$$

Problem $\mathcal{P}7$ is a non-convex optimization problem due to non-convex objective function. To reduce required computational capacity to solve $\mathcal{P}7$, $\mathcal{P}7$ is transformed to a convex optimization problem by decomposing the objective function and introducing auxiliary variables as follows.

$$\mathcal{P}8: \min_{P,x} \quad \sum_{u \in \mathcal{U}^{(c)}} \sum_{k \in \mathcal{K}} \frac{\Delta f}{2} \log_2 \left(1 + \sum_{v \in \mathcal{U}^{(c)} \setminus u} (\sigma_{vk}^{(u)})^2 P_{vk}\right) - x_{uk}$$

s.t.    C14, C15

$$C16: \frac{\Delta f}{2} \log_2 \left(1 + 2\sigma_{uk}^2 P_{uk} + \sum_{v \in \mathcal{U}^{(c)} \setminus u} (\sigma_{vk}^{(u)})^2 P_{vk}\right) \geq x_{uk},$$

$$\forall u \in \mathcal{U}^{(c)}, k \in \mathcal{K},$$

where $x$ is a vector consisting all $x_{uk}$ variables. $\mathcal{P}8$ is a concave minimization problem, and it can be optimally solved using Algorithm 5 [87], where $\delta$ is the error tolerance.


### 5.5.3   Implementation of Algorithm 5

Implementation of several steps of Algorithm 5 is discussed in this section.

---

**Algorithm 5** : Subcarrier Allocation and Water-Level Calculation

---

1: Let $O(P,x)$ denote the objective function;
2: Let $F$ denotes the feasible set of $\{P,x\}$ which satisfies constraints C14–C16;
3: Find a feasible solution $\{P_f, x_f\}$ which satisfies C14–C16. E.g., $\{0, \ldots, 0, 0, \ldots, 0\}$;
4: Find a linear polyhedron $F^0$ which encloses $F$;
5: Find the vertices $(V^0)$ of $F^0$;
6: $s \leftarrow 1$;
7: **while** Optimal is not found **do**
8:     Choose the vertex $\{P_m, x_m\}$ which minimizes $O(P,x)$, where $\{P_m, x_m\} \in V^{s-1}$;
9:     Find the smallest $\lambda$ $(0 \leq \lambda \leq 1)$ such that $\{\lambda P_f + (1-\lambda)P_m, \lambda x_f + (1-\lambda)x_m\} \in F$;
10:     $\{P_s, x_s\} \leftarrow \{\lambda P_f + (1-\lambda)P_m, \lambda x_f + (1-\lambda)x_m\}$;
11:     **if** $O(P_m, x_m) - O(P_s, x_s) \leq \delta$ **then**
12:         $\{P_s, x_s\}$ is optimal;
13:     **else**
14:         Find an active constraint at $\{P_s, x_s\}$, that is $C^s(P_s, x_s) = 0$, where $C^s(P,x) \in \{C14, C15, C16\}$;
15:         Generate a new constraint $[\nabla C^s(P_s, x_s)](\{P,x\} - \{P_s, x_s\}) \leq 0$;
16:         Add the new constraint to the constraint set $F^{s-1}$, and find the new vertices due to this constraint. Form $V^s$ by adding the new vertices to $V^{s-1}$;
17:         $s \leftarrow s + 1$;
18:     **end if**
19: **end while**

---

### 5.5.3.1  Steps 4 and 5

First, a tight upper bound $\theta$ for $\sum_{u \in \mathcal{U}^{(c)}} \sum_{k \in \mathcal{K}} (P_{uk} + x_{uk})$ is determined subject to C14–C16. As $x \geq \ln(1 + x)$, $\theta = \beta \Delta f / \ln(4) + \sum_{u \in \mathcal{U}^{(c)}} P_{T,u}$, and $\beta$ is found by solving

$$\mathcal{P}9 : \beta = \max_{P} \quad \sum_{u \in \mathcal{U}^{(c)}} \sum_{k \in \mathcal{K}} \left( 2\sigma_{uk}^2 P_{uk} + \sum_{v \in \mathcal{U}^{(c)} \setminus u} (\sigma_{vk}^{(u)})^2 P_{vk} \right)$$

$$\text{s.t.} \quad \text{C14, C15.}$$

Problem $\mathcal{P}9$ is a linear programming problem, and its objective function is derived from C16 using relationship of $x \geq \ln(1 + x)$. Then, $\mathcal{P}9$ is rewritten by rearranging its objective function as follows.

$$\mathcal{P}10 : \beta = \max_{P} \quad \sum_{u \in \mathcal{U}^{(c)}} \sum_{k \in \mathcal{K}} P_{uk} \left( 2\sigma_{uk}^2 + \sum_{v \in \mathcal{U}^{(c)} \setminus u} (\sigma_{uk}^{(v)})^2 \right)$$

$$\text{s.t.} \quad \text{C14, C15.}$$

The optimal value of $P_{uk}, \forall u, k$ for $\mathcal{P}10$, hence for $\mathcal{P}9$, is given by

$$P_{uk} = \begin{cases} P_{T,u}, & \text{if } k = \arg \max_{\forall k \in \mathcal{K}} \{ 2\sigma_{uk}^2 + \sum_{v \in \mathcal{U}^{(c)} \setminus u} (\sigma_{uk}^{(v)})^2 \}; \\ 0, & \text{otherwise.} \end{cases} \tag{5.14}$$

Therefore,

$$\beta = \sum_{u \in \mathcal{U}^{(c)}} P_{T,u} \max_{\forall k \in \mathcal{K}} \{ 2\sigma_{uk}^2 + \sum_{v \in \mathcal{U}^{(c)} \setminus u} (\sigma_{uk}^{(v)})^2 \}. \tag{5.15}$$

Thus, the initial linear polyhedron $F^0$ is given by $\{ P, x \mid \sum_{u \in \mathcal{U}^{(c)}} \sum_{k \in \mathcal{K}} P_{uk} + x_{uk} \leq \theta \}$.

The initial set of vertices $V^0$ consists of $2|\mathcal{K}||\mathcal{U}^{(c)}|$ vertices with each vertex being represented by $|\mathcal{K}||\mathcal{U}^{(c)}|$ of $P_{uk}$ and $|\mathcal{K}||\mathcal{U}^{(c)}|$ of $x_{uk}$ variables, where $|\mathcal{Y}|$ denotes the number of elements in $\mathcal{Y}$. The $i$th vertex's $i$th variable equals to $\theta$ while all other variables of the vertex are zeros. Only the vertices which produce better objective function values than the initial feasible point are required to be stored and considered in the algorithm.

### 5.5.3.2  Step 9

The smallest value of $\lambda$ is found using a bisection algorithm. In the bisection algorithm, $\lambda$ is decreased when all constraints are over satisfied, and $\lambda$ is increased when there are constraints that are not satisfied.

### 5.5.3.3  Steps 14 and 15

Vertex $\{P_m, x_m\}$ is eliminated from $V^s$ using an additional constraint. First, select an active constraint $C^s(P, x)$ at $\{P_s, x_s\}$, i.e., $C^s(P_s, x_s) = 0$. Second, add a new constraint as follows. If $C^s(P, x) \in$ C14 and corresponds to the $u$th user, the new constraint is

$$\sum_{k \in \mathcal{K}} P_{uk} \leq \sum_{k \in \mathcal{K}} P_{uk}^{(s)}, \tag{5.16}$$

where $P_{uk}^{(s)} \in P_s$. Otherwise, if $C^s(P, x) \in$ C16 and corresponds to the $u$th user and the $k$th subcarrier, the new constraint is

$$x_{uk} - \frac{\Delta f}{\ln(4) 2^{(2x_{uk}^{(s)}/\Delta f)}} \left( 2\sigma_{uk}^2 P_{uk} + \sum_{v \in \mathcal{U}^{(c)} \setminus u} (\sigma_{vk}^{(u)})^2 P_{vk} \right) \leq x_{uk}^{(s)} + \frac{\Delta f}{\ln(4)} \left( 1 - \frac{1}{2^{(2x_{uk}^{(s)}/\Delta f)}} \right), \tag{5.17}$$

where $x_{uk}^{(s)} \in x_s$.

### 5.5.3.4  Step 16

First, create a simplex tableau using coefficients of inequalities in $F^{s-1}$ and non-zero variables of $\{P_m, x_m\}$ as the basic variables. Second, add the new constraint to the simplex tableau as a new row. Finally, find all the new vertices, which are to be added to $V^s$, by performing dual pivot operations on all the non-basic variables.

## 5.6  Performance Evaluation

Performance of the proposed resource allocation scheme is evaluated using a cluster consisting of 15 small cell BSs and three macro cell BSs. Small cell BSs are uniformly and randomly distributed over a $100\,m \times 100\,m$ area at the center of the cluster. Macro cell BSs are located $300\,m$ from the center of the cluster, similar to the network shown in Fig. 5.1. Radiuses of a small cell and a macro cell are $30\,m$ and $1000\,m$, respectively. Total bandwidth of $20\,MHz$ is divided into 128 subcarriers. $|\mathcal{U}^{(c)}|$ number of users are uniformly distributed over the $100\,m \times 100\,m$ area. Channels between the BSs and the users are modeled as Rayleigh faded channels with path loss being proportional to $x^{-\eta}$, where $\eta$ is the path loss exponent and $x$ is the distance between the user and the BS. Total delay when access the cloud (i.e., $d_{Total}$) is 594 ms [84]; thus, $T_S$ should be at least 594 ms. $T_F$ is 1 ms, similar to the resource allocation interval in the LTE standard. Remaining parameters are shown in Table 5.1.

**Table 5.1** Simulation parameters

| Parameter | Value (unit) |
|---|---|
| Single sided power spectra density of noise, $N_0$ | $-174\,\text{dBm/Hz}$ |
| Path-loss exponent, $\eta$ | 4 |
| Error tolerance in Algorithm 5, $\delta$ | 0.5% |
| Transmission delay, $d_t$ | 7 ms |
| Queuing delay, $d_q$ | 400 ms |
| Processing time, $d_p$ | 180 ms |
| $l_b$ for $d_t$ calculation | 200 km |
| $\xi$ for $d_q$ calculation | $5\,\text{s}^{-1}$ |
| $1/\varphi_s$ for $d_q$ calculation | $1/10\,\text{s}^{-1}$ |
| $\chi_s^2$ for $d_q$ calculation | 0.05 |
| $\alpha_s$ for $d_p$ calculation | 100 ms |
| $\nu_s$ for $d_p$ calculation | 80 ms |

Benchmark resource allocation scheme used for performance comparison allocates resources as follows. First, it uses FFR to determine the subcarrier subsets that users can use without causing CCI. Second, it optimally allocates subcarriers and transmit power within each cell by solving a joint resource allocation problem. Use of FFR is beneficial as it can be implemented without causing a significant signaling overhead in backhauls or requiring small cell BSs to perform complex computations. With FFR, users lie within 15 m from a small cell BS are able to access all 128 subcarriers while users lie beyond are able to access only a subset of the subcarriers. Number of subcarriers in one of these subsets equals to 128 divided by the number of overlapping cell coverages. Users who are outside small cell coverages are allocated to the macro cell BSs. In macro cells, users lie within 150 m from a macro cell BS are able to access all the subcarriers while the remaining users are able to access only a subset of the subcarriers. Once FFR determines the subcarrier subsets for different areas of each macro and small cell, subcarriers and transmit power levels are optimally allocated within each cell. For this purpose, Algorithms 1 and 2 derived in Chap. 3 are used with the following three modifications: (1) upper and lower level time slot durations are set to be equal to $T_F$; (2) WLAN resources and QoS constraints are ignored; and (3) if FFR allows the $u$th user to access only the subcarriers in subset $\mathcal{K}_u$, then $\rho_{u,k}^{C*} = 0, \forall k \notin \mathcal{K}_u$ in Algorithm 2.

Throughput performance of the proposed and the benchmark schemes is compared in Fig. 5.4. The proposed scheme achieves higher throughput performance due to two reasons. First, it jointly allocates subcarriers and transmit power considering CCI in the entire cluster. Second, it increases frequency reuse by allowing all the users to access all the subcarriers. In the benchmark, frequency reuse is reduced as cell-edge users are able to access only a subset of the available subcarriers. Furthermore, when $T_S$ increases from 594 to 1188 ms, throughput achieved by

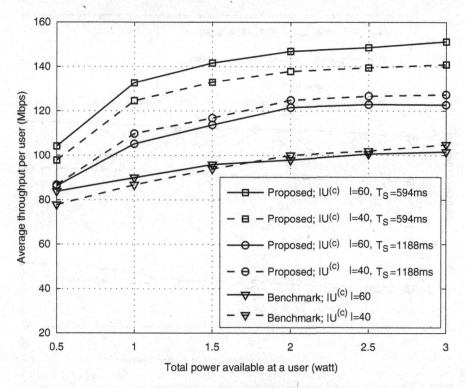

**Fig. 5.4** Throughput performance of the proposed and the benchmark schemes

the proposed scheme reduces, because effectiveness of the resource allocation decisions made at the cloud decreases with $T_S$ as the channel gains change over time.

Complexities of the proposed and the benchmark schemes are compared in Fig. 5.5. Complexity of the proposed scheme is measured in terms of the number of required iterations for the *while* loop in step 7 of Algorithm 5. Complexity of the benchmark is measured in terms of the number of required iterations for the inner most loop. Complexity of the proposed scheme is higher than that of the benchmark as the proposed scheme jointly allocates subcarriers for the entire cluster, whereas the benchmark jointly allocates subcarriers for a cell only. However, Algorithm 5 is executed at the cloud. Therefore, the proposed scheme only requires the small cell BSs to calculate the user transmit power levels using (5.8). In the proposed scheme, the complexity slightly increases with $P_{T,u}$. When $P_{T,u}$ increases, the feasible region enlarges, and there are more vertices on the boundary of the feasible region. As more vertices should be checked for optimality, the complexity of the scheme increases.

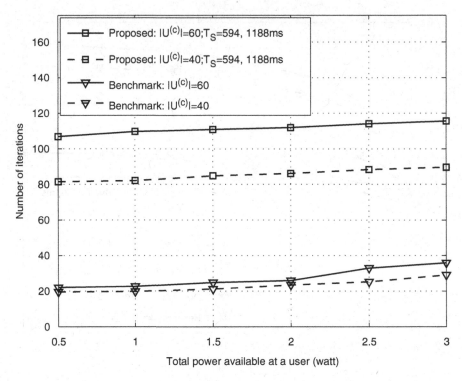

**Fig. 5.5** Complexities of the proposed and the benchmark schemes

## 5.7  Summary

A resource allocation scheme to allocate users, subcarriers and transmit power for interworking macro cell and hyper-dense small cell networks is proposed in this chapter. The proposed scheme uses cloud computing to reduce computational burden for low-cost small cell BSs. It operates on two time-scales to overcome the effect of high delay when access the cloud and to reduce signaling overhead in BS backhauls. In the slow time-scale, resource allocation decisions are made at the cloud. In the fast time-scale, BSs further adapt these decisions to fast varying wireless channel conditions. Simulation results demonstrate that the proposed scheme provides better throughput performance than the benchmark scheme, and its performance increases as the duration of a slow time-scale time slot decreases.

# Chapter 6
# Conclusions and Future Directions

This book investigates resource allocation for heterogeneous wireless networks which employ interworking, D2D communication and hyper-dense small cell deployments to enhance the network throughput, QoS support and coverage. In this chapter, the key conclusions made based on this investigation are summarized and the future research directions are discussed.

## 6.1 Conclusions

Resource allocation for interworking cellular networks and WLANs is investigated in Chap. 3. The key challenge for allocating resources for this interworking system is the high complexity of the resource allocation algorithms due to following two reasons: (1) resources have to be allocated based on multiple PHY and MAC technologies of the two networks, and (2) cellular networks and WLANs have different time-scales for resource allocation. To allocate resources overcoming this challenge, a MMDP based and a low time-complex heuristic schemes are proposed. These schemes operate on two time-scales, jointly allocate resources of the two networks based on the PHY and MAC technologies, and allow users to multi-home. They provide higher throughput and better QoS performance compared to that of the benchmarks. Therefore, the performance of the interworking system can be improved by jointly allocating resources of the two networks while allowing the users to multi-home.

A systematic approach to design resource allocation schemes that allocate resources based on multiple PHY and MAC technologies is presented in Chap. 3. In this approach, a novel mechanism to allocate uplink transmit power of WLAN users who use a MAC similar to DCF or HCF is presented. Such power allocation is particularly useful to efficiently allocate power of multi-homing users in interworking systems. Furthermore, the concept of allocating resources in two different

81
A.T. Gamage, X.(S.) Shen, *Resource Management for Heterogeneous Wireless Networks*, SpringerBriefs in Electrical and Computer Engineering, DOI 10.1007/978-3-319-64268-0_6

time-scales is also demonstrated. The reason for using two time-scales is that the cellular network and the WLAN have resource allocation intervals with different durations. The multiple time-scale approach can also be used in the scenarios where the execution time of an algorithm is longer than the decision execution interval. In this scenario, the algorithm can be divided into two sub-algorithms. The first sub-algorithm has a longer execution time, operates in a slow time-scale, and makes long-term decisions. The second sub-algorithm has an execution time shorter than the decision execution interval, operates in a fast time-scale, and optimizes the long-term decisions to current conditions of the system. The effectiveness of using two time-scales in this scenario is validated in Chap. 5.

Resource allocation for D2D communication underlaying cellular/WLAN inter-working system is investigated in Chap. 4. There are two main advantages to integrating D2D communication with interworking. First, D2D communication can improve network throughput and QoS performance when interworking cannot. Second, high capacity, secure and non-interfering WLAN-based D2D links can be setup among the users in proximity. These users do not have to be within a WLAN AP coverage. To setup these links, control and authentication signals are sent through the cellular network as users can multi-home. However, the high complexity and high signaling overhead caused by mode selection and interference management bring significant challenges for resource allocation when D2D communi-cation is integrated. To overcome these challenges, a resource allocation scheme, which operates on three time-scales and allocates orthogonal and non-orthogonal resources via two steps, is proposed. Using a slow time-scale for mode selection, average complexity and average signaling overhead are reduced. Using two steps to allocate orthogonal and non-orthogonal resources, interference management is simplified. Simulation results demonstrate that the proposed scheme provides significant throughput and QoS performance enhancements compared to that of the benchmarks. Therefore, the performance of the interworking system can be further improved by integrating D2D communication. The complexity and the signaling overhead can be reduced by operating the resource allocation scheme over multiple time-scales.

Resource allocation for interworking macro cell and hyper-dense small cell networks is studied in Chap. 5. The main challenge for allocating resources is the high complexity of the resource allocation schemes. The high complexity is a result of jointly allocating a large number of resources subject to a large number of CCI constraints. To overcome this challenge, a resource allocation scheme which operates on two time-scales using cloud computing is proposed. The complex part of the scheme is executed at the cloud over a slow time-scale. It is executed over a slow time-scale as the time duration takes to send channel information to the cloud, determine resource allocation decisions at the cloud and redistribute these decisions to the BSs is longer than the resource allocation interval of the cellular networks. At the BSs, these decisions are re-optimized based on the instantaneous channel gains. Simulation results show that the proposed scheme provides higher throughput than that of the benchmark. Therefore, resources of this interworking system or any other large-scale network can be jointly allocated using cloud computing to improve throughput. The negative effect of high delay when access the cloud can be overcome by operating the resource allocation scheme over two time-scales.

## 6.2  Future Research Directions

Future research directions to extend the resource allocation scenarios investigated in this book include, but not limited to, the following three areas: (1) low-complex techniques to allocate resources over multiple time-scales, (2) efficient clustering techniques for interworking macro cell and hyper-sense small cell networks, and (3) resource allocation schemes for interworking macro cell and hyper-sense small cell networks imposing QoS and backhaul constraints.

Investigation of techniques that simplify allocating resources over multiple time-scales is a one of the most important future research directions, as the resource allocation schemes which operate over multiple time-scales have a wide range of applications. Three of these applications are demonstrated in Chaps. 3, 4 and 5. Two techniques investigated in this book are as follows. First, an optimal MMDP based resource allocation scheme to allocate resources over multiple time-scales is investigated in Chap. 3. This scheme has a high complexity due to existence of a large state space. Second, a resource allocation framework that uses average throughputs and average transmit power levels determined using time-correlated wireless channels is investigated in Chap. 5. This framework eliminates the complexity of the MMDP based scheme. However, it uses complex expressions, which cannot be directly used in resource allocation schemes, to calculate average throughputs and average transmit power levels. Therefore, interested readers are encouraged to further simplify these expressions and derive efficient resource allocation schemes based on this framework.

Another important future research direction is the investigation of clustering techniques that divide interworking macro cell and hyper-dense small cell networks into clusters. Investigation of clustering techniques is important due to two main reasons. First, jointly allocating resources to the entire interworking system is not practical as there are a large number of resources in the system. Due to this reason, in Chap. 5, the interworking system is assumed to be clustered. Second, performance of the interworking system depends on how the system is clustered. The performance of the system increases with the cluster size as more resources are jointly allocated. However, increasing the cluster size increases the required time duration to make resource allocation decisions, due to the increased problem size. Thus, as the cluster size increases, effectiveness of the resource allocation decisions decrease. Consequently, the performance of the interworking system reduces as well. Therefore, it is crucial to investigate efficient clustering techniques that maximize the interworking system performance.

Investigation of allocating resources to interworking macro cell and hyper-sense small cell networks imposing QoS and backhaul constraints is an important research direction. Considering these constraints during resource allocation is important as a majority of services have QoS requirements and that small cell BS backhauls have limited capacities. In Chap. 5, these constraints are not considered as the objective is to gain valuable insights on the effectiveness of the proposed cloud assisted two time-scale resource allocation scheme for this interworking system. The interested readers are encouraged to investigate allocating resources considering these constraints.

# Appendix A

## A.1 Proof of Convexity of $R_i^{CB}(\mathbf{P}^{CB})$

Convexity of $R_i^{CB}(\mathbf{P}^{CB})$ given by (2.5) is proved as follows. Let

$$f(\mathbf{x}) = \frac{-1}{k_0 + \sum_{i=1}^{N} \frac{k_i}{x_i}}, \tag{A.1}$$

where $k_i \in \mathbb{R}^+, \forall i \in \{0, \ldots, N\}$; $\mathbf{x} = [x_1, \ldots, x_N]$; and $x_i \in \mathbb{R}^+, \forall i \in \{1, \ldots, N\}$. Next, consider

$$f(\mathbf{z}) - f(\mathbf{x}) - \nabla f(\mathbf{x})[\mathbf{z} - \mathbf{x}]^T = \left( \frac{1}{k_0 + \sum_{i=1}^{N} \frac{k_i}{z_i}} \right) \left( \frac{1}{k_0 + \sum_{i=1}^{N} \frac{k_i}{x_i}} \right)^2 g(\mathbf{x}), \tag{A.2}$$

where

$$g(\mathbf{x}) = \left( k_0 + \sum_{i=1}^{N} \frac{k_i}{z_i} \right) \left( k_0 + \sum_{i=1}^{N} \frac{k_i z_i}{x_i^2} \right) - \left( k_0 + \sum_{i=1}^{N} \frac{k_i}{x_i} \right)^2 \tag{A.3}$$

© The Author(s) 2018

A.T. Gamage, X.(S.) Shen, *Resource Management for Heterogeneous Wireless Networks*, SpringerBriefs in Electrical and Computer Engineering, DOI 10.1007/978-3-319-64268-0

and $\mathbf{z} = [z_1, \ldots, z_N]$ with $z_i \in \mathbb{R}^+$, $\forall i \in \{1, \ldots, N\}$. Furthermore, $g(\mathbf{x})$ can also be written as

$$
g(\mathbf{x}) = k_0 \sum_{i=1}^{N} k_i \left( \frac{1}{\sqrt{z_i}} - \frac{\sqrt{z_i}}{x_i} \right)^2 + \sum_{i=2}^{N} \sum_{j=1}^{\lfloor i/2 \rfloor} k_{i-j+1} k_j \left( \sqrt{\frac{z_j}{z_{i-j+1}x_j^2}} - \sqrt{\frac{z_{i-j+1}}{z_j x_{i-j+1}^2}} \right)^2
$$

$$
+ \sum_{j=2}^{N-1} \sum_{i=1}^{\lfloor j/2 \rfloor} \left[ k_{N-i+1} k_{N-j+i} \times \left( \sqrt{\frac{z_{N-j+i}}{z_{N-i+1}x_{N-j+i}^2}} - \sqrt{\frac{z_{N-i+1}}{z_{N-j+i}x_{N-i+1}^2}} \right)^2 \right].
$$

(A.4)

Since $k_i \in \mathbb{R}^+$, $\forall i \in \{0, \ldots, N\}$, $g(\mathbf{x}) \geq 0$. Therefore, from (A.2), $f(\mathbf{x})$ is a convex function as it satisfies the first order condition [50]. Furthermore, $f(\mathbf{x})$ is also a non-increasing function as

$$
\frac{\partial f(\mathbf{x})}{\partial x_i} = \frac{-k_i}{x_i^2} \left( \frac{1}{k_0 + \sum_{i=1}^{N} \frac{k_i}{x_i}} \right)^2 \leq 0, \ \forall i \in \{1, \ldots, N\}.
$$

(A.5)

Now, substitute $x_i = \log_2(1 + y_i)$, $\forall i \in \{1, \ldots, N\}$ to (A.1). Then,

$$
f(\mathbf{y}) = \frac{-1}{k_0 + \sum_{i=1}^{N} \frac{k_i}{\log_2(1+y_i)}}.
$$

(A.6)

As $x_i = \log_2(1 + y_i)$ is a concave function with respect to $y_i \in \mathbb{R}^+$, $f(\mathbf{x})$ is a convex function and $f(\mathbf{x})$ is also non-increasing in each of its argument, from vector composition theory [50], $f(\mathbf{y})$ given by (A.6) is a convex function. Therefore, $R_i^{CB}(\mathbf{P}^{CB})$ given by (2.5) is a concave function.

## A.2 Proof of Convexity of Constraint C4

Convexity of the constraint C4 in Sect. 3.5.2 is proved as follows. From (2.7), the derivative of $P_{avg,i}^{CB}(\mathbf{P}^{CB})$ when $P_i^{CB} > 0$ is given by

$$
\frac{\partial P_{avg,i}^{CB}(\mathbf{P}^{CB})}{\partial P_i^{CB}} = \frac{T_{CP}}{T_P B^W \cdot \left( \log_2(1 + P_i^{CB}\alpha_i^W) \right)^2} \times \left[ R_i^{CB}(\mathbf{P}^{CB}) \left( \log_2(1 + P_i^{CB}\alpha_i^W) \right. \right.
$$

$$
\left. \left. - \frac{P_i^{CB}\alpha_i^W}{\ln(2)(1 + P_i^{CB}\alpha_i^W)} \right) + \frac{\partial R_i^{CB}(\mathbf{P}^{CB})}{\partial P_i^{CB}} P_i^{CB} \log_2(1 + P_i^{CB}\alpha_i^W) \right].
$$

(A.7)

Next, let

$$
g(x) = \log_2(1 + x) - \frac{x}{(1 + x)\ln(2)}, \ x \geq 0.
$$

(A.8)

Function $g(x) \geq 0$ as $g(0) = 0$ and $dg(x)/dx \geq 0, \forall x \in \mathbb{R}^+$. Also, $R_i^{CB}(\mathbf{P}^{CB})$ is a positive non-decreasing concave function. Therefore, from (A.7), $\partial P_{avg,i}^{CB}(\mathbf{P}^{CB})/\partial P_i^{CB} \geq 0$. Thus, $P_{avg,i}^{CB}(\mathbf{P}^{CB})$ is a non-decreasing function of $P_i^{CB}$. Hence, $P_{avg,i}^{CB}(\mathbf{P}^{CB}) \leq P_{T,i}$ is a convex set [50]. Therefore, $\{P_{avg,i}^C, \bar{P}_{i,j}^{CF}, P_i^{CB} | \text{C4 is satisfied}, i \in \mathcal{S}_N, j \in \mathcal{K}^{CF}\}$ is a convex set. That is, C4 is a convex constraint. Moreover, $\{\bar{P}_{i,k}^C, \bar{P}_{i,j}^{CF}, P_i^{CB} | \text{C4 is satisfied}, i \in \mathcal{S}_N, j \in \mathcal{K}^{CF}, k \in \mathcal{K}^C\}$ is also a convex set as $P_{avg,i}^C$ is a linear combination of $\bar{P}_{i,k}^C, \forall k$ (see (3.10)).

## A.3   Proof of Existence of a Solution for (3.24)

Existence of a solution for (3.24) is shown as follows. By differentiating both sides of (2.7) with respect to $P_i^{CB}$, and then removing the non-negative term,

$$\frac{\partial \bar{R}_i^{CB}(\mathbf{P}^{CB})}{\partial P_i^{CB}} \begin{cases} = \frac{B^W \alpha_i^W}{\ln(2)} \cdot \frac{\partial P_{avg,i}^{CB}(\mathbf{P}^{CB})}{\partial P_i^{CB}}, \text{ at } P_i^{CB} = 0; \\ < \frac{B^W \log_2(1+\alpha_i^W P_i^{CB})}{P_i^{CB}} \cdot \frac{\partial P_{avg,i}^{CB}(\mathbf{P}^{CB})}{\partial P_i^{CB}}, \text{ for } P_i^{CB} > 0. \end{cases} \tag{A.9}$$

Derivatives $\partial \bar{R}_i^{CB}(\mathbf{P}^{CB})/\partial P_i^{CB}$ and $\partial P_{avg,i}^{CB}(\mathbf{P}^{CB})/\partial P_i^{CB}$ are positive and monotonically decreasing functions of $P_i^{CB}$ as $\bar{R}_i^{CB}(\mathbf{P}^{CB})$ and $P_{avg,i}^{CB}(\mathbf{P}^{CB})$ are concave increasing functions of $P_i^{CB}$. Also, the value of $(B^W \log_2(1 + \alpha_i^W P_i^{CB}))/P_i^{CB}$ decreases from $B^W \alpha_i^W/\ln(2)$ to 0 as $P_i^{CB}$ goes from 0 to $\infty$. Thus, there exist a $p_i, (p_i > 0)$ such that

$$\frac{\partial \bar{R}_i^{CB}(\mathbf{P}^{CB})}{\partial P_i^{CB}} < \frac{\mu_i^*}{1 + \lambda_i} \cdot \frac{\partial P_{avg,i}^{CB}(\mathbf{P}^{CB})}{\partial P_i^{CB}}, \text{ for } P_i^{CB} > p_i. \tag{A.10}$$

Therefore, from (3.22), (3.23) and (A.10), there exist a $P_i^{CB*}, (P_i^{CB*} \in (0, p_i))$ which is the solution for (3.24).

## A.4   Proof of Convergence of the Algorithm Which Calculates $\mathbf{P}^{CB*}$

Convergence of the iterative algorithm which calculates $\mathbf{P}^{CB*}$ is proved as follows. At the $n$th iteration, $P_i^{CB*}$ is calculated using (3.24) such that the left and right hand sides of (A.10) are equal (see the derivation of (3.24)). Since

$$\frac{\left(\frac{\partial^2 \bar{R}_i^{CB}(\mathbf{P}^{CB})}{\partial P_j^{CB} \partial P_i^{CB}}\right)}{\left(\frac{\partial \bar{R}_i^{CB}(\mathbf{P}^{CB})}{\partial P_i^{CB}}\right)} > \frac{\left(\frac{\partial^2 P_{avg,i}^{CB}(\mathbf{P}^{CB})}{\partial P_j^{CB} \partial P_i^{CB}}\right)}{\left(\frac{\partial P_{avg,i}^{CB}(\mathbf{P}^{CB})}{\partial P_i^{CB}}\right)}, \forall P_j^{CB} > 0, i \neq j, \tag{A.11}$$

if $P_j^{CB*}$ increases at the $n$th iteration, the left hand side of (A.10) becomes larger than the right hand side of (A.10). Therefore, $P_i^{CB*}$ increases at the $(n+1)$th iteration. Thus, by induction, $P_i^{CB*}$, $\forall i$ increase in each iteration. However, $P_i^{CB*}$ is upper bounded by the $p_i$ which satisfies

$$\frac{B^W \log_2(1 + \alpha_i^W p_i)}{p_i} = \frac{\mu_i^*}{1 + \lambda_i}. \tag{A.12}$$

Thus, the iterative algorithm converges. Furthermore, $P_i^{CB*}$ converges to a value which is smaller than $p_i$; otherwise (i.e., if $P_i^{CB*} = p_i$), (A.9) does not hold true.

# Appendix B

## B.1 Calculation of Average User Transmit Power

Average transmit power of the $u$th user during the next slow time-scale time slot over the $k$th subcarrier is given by

$$P_{uk} = \mathbb{E}\left\{ \frac{1}{L} \sum_{t=0}^{L-1} p_{ukt}(H_{ukt}) \Big| H_{uk(L_0-L)} \right\}$$

$$= \frac{1}{L} \sum_{t=0}^{L-1} \mathbb{E}\left\{ p_{ukt}(H_{ukt}) \Big| H_{uk(L_0-L)} \right\}, \tag{B.1}$$

where $H_{uk(L_0-L)}$ is the CSI transmitted to the cloud at the $L_0$th fast time-scale time slot within the current slow time-scale time slot. Furthermore, $Pr(H_{ukt}|H_{uk(L_0-L)})$ is given by (5.6) with $\bar{\sigma}^2 = (1 - \rho^{2(t+L-L_0)})\sigma_{uk}^2/2$ and $s = \rho^{(t+L-L_0)}\sqrt{H_{uk(L_0-L)}}$, where $\sigma_{uk}^2$ is the average normalized power gain of the channel between the $u$th user and the BS to which the $u$th user is connected to, over the $k$th subcarrier. Normalized power gain is the ratio of channel power gain to noise power. From (5.6), (5.7) and (5.8),

© The Author(s) 2018
A.T. Gamage, X.(S.) Shen, *Resource Management for Heterogeneous Wireless Networks*, SpringerBriefs in Electrical and Computer Engineering, DOI 10.1007/978-3-319-64268-0

$$\mathbb{E}\{p_{ukt}(H_{ukt})|H_{uk(L_0-L)}\} = \int_0^\infty \left[\mu_{uk} - \frac{1+I_{uk}}{y}\right]^+ Pr(H_{ukt} = y|H_{uk(L_0-L)})dy$$

$$= \int_{(1+I_{uk})/\mu_{uk}}^\infty \left(\mu_{uk} - \frac{1+I_{uk}}{y}\right) Pr(H_{ukt} = y|H_{uk(L_0-L)})dy$$

$$= \int_{(1+I_{uk})/\mu_{uk}}^\infty \frac{1}{2\bar{\sigma}^2}\left(\mu_{uk} - \frac{1+I_{uk}}{y}\right) e^{-(s^2+y)/2\bar{\sigma}^2}$$

$$\sum_{k=0}^\infty \frac{y^k(s/2\bar{\sigma}^2)^{2k}}{(k!)^2} dy.$$

$$(B.2)$$

By substituting $x^2 = y/\bar{\sigma}^2$, (B.2) can be written as

$$\mathbb{E}\{p_{ukt}(H_{ukt})|H_{uk(L_0-L)}\} = \int_b^\infty \left(\mu_{uk}x - \frac{1+I_{uk}}{\bar{\sigma}^2 x}\right) e^{-(x^2+a^2)/2} I_0(ax)dx, \qquad (B.3)$$

where $a = s/\bar{\sigma}$, $b = \sqrt{(1+I_{uk})/(\bar{\sigma}^2\mu_{uk})}$, and $I_v(\theta)$ is the $v$th order modified Bessel function of first kind which is defined as [85]

$$I_v(\theta) = \sum_{k=0}^\infty \frac{(\theta/2)^{2k+v}}{k!(k+v)!}, \quad \forall v \in \mathbb{Z}^+. \qquad (B.4)$$

Furthermore, the 1st order generalized Marcum Q-function is given by Proakis [85]

$$Q_1(\alpha, \beta) = \int_\beta^\infty x e^{-(x^2+\alpha^2)/2} I_0(\alpha x)dx$$

$$(B.5)$$

$$= e^{-(\alpha^2+\beta^2)/2} \sum_{k=0}^\infty \left(\frac{\alpha}{\beta}\right)^k I_k(\alpha\beta).$$

By (B.3)–(B.5),

$$
\mathbb{E}\{p_{ukt}(H_{ukt})|H_{uk(L_0-L)}\} = \mu_{uk}Q_1(a,b) - \frac{1+I_{uk}}{\bar{\sigma}^2}\int_b^\infty \frac{1}{x}e^{-(x^2+a^2)/2}I_0(ax)dx.
$$

(B.6)

Next, the second term on the right hand side of (B.6) is integrated as follows.

$$
\int_b^\infty \frac{1}{x}e^{-(x^2+a^2)/2}I_0(ax)dx = \int_b^\infty \frac{1}{x}e^{-(x^2+a^2)/2}dx + \int_b^\infty \frac{1}{x}e^{-(x^2+a^2)/2}\sum_{k=1}^\infty \frac{(ax/2)^{2k}}{(k!)^2}dx
$$

$$
= 0.5e^{-a^2/2}E_1(b^2/2)
$$

$$
- \int_b^\infty \frac{d\left(e^{-(x^2+a^2)/2}\right)}{dx}\sum_{k=1}^\infty \frac{a^2}{4}\frac{(ax/2)^{2k-2}}{(k!)^2}dx
$$

$$
= 0.5e^{-a^2/2}E_1(b^2/2) + e^{-(a^2+b^2)/2}\sum_{k=0}^\infty \frac{a^2}{4}\frac{(ab/2)^{2k}}{((k+1)!)^2}
$$

$$
+ \int_b^\infty e^{-(x^2+a^2)/2}\sum_{k=2}^\infty \frac{(k-1)a^3}{4}\frac{(ax/2)^{2k-3}}{(k!)^2}dx,
$$

(B.7)

where $E_1(\theta)$ is the exponential integral given by (3.30), [86]. There are tight closed form approximations for $E_1(\theta)$ [88]. By further integrating (B.7) using integration by parts,

$$
\int_b^\infty \frac{1}{x}e^{-(x^2+a^2)/2}I_0(ax)dx = 0.5e^{-a^2/2}E_1(b^2/2)
$$

$$
+ e^{-(a^2+b^2)/2}\sum_{i=0}^\infty\sum_{k=0}^\infty \frac{a^{2i+2}}{2^{i+2}}\frac{(k+i)!(ab/2)^{2k}}{k!((k+i+1)!)^2}.
$$

(B.8)

Therefore, by substituting (B.8) into (B.6),

$$\mathbb{E}\{p_{ukt}(H_{ukt})|H_{uk(L_0-L)}\}$$

$$= \mu_{uk}Q_1(a,b) - \frac{1+I_{uk}}{\bar{\sigma}^2}\left[0.5e^{-a^2/2}\mathrm{E}_1(b^2/2)\right.$$

$$\left.+ e^{-(a^2+b^2)/2}\sum_{i=0}^{\infty}\sum_{k=0}^{\infty}\frac{a^{2i+2}}{2^{i+2}}\frac{(k+i)!(ab/2)^{2k}}{k!((k+i+1)!)^2}\right]$$

$$= -\frac{1+I_{uk}}{2\bar{\sigma}^2}e^{-a^2/2}\mathrm{E}_1(b^2/2) \qquad\qquad\qquad\qquad (\mathrm{B.9})$$

$$+ e^{-(a^2+b^2)/2}\sum_{i=0}^{\infty}\sum_{k=0}^{\infty}\frac{a^{2i}}{2^ik!}\left(\frac{ab}{2}\right)^{2k}\left[\frac{\mu_{uk}}{(k+i)!} - \frac{(k+i)!(1+I_{uk})a^2}{((k+i+1)!)^24\bar{\sigma}^2}\right]$$

$$= -\frac{1+I_{uk}}{2\bar{\sigma}^2}e^{-a^2/2}\mathrm{E}_1(b^2/2)$$

$$+ e^{-(a^2+b^2)/2}\sum_{i=0}^{\infty}\sum_{k=0}^{\infty}\frac{(a^2/2)^i(ab/2)^{2k}}{k!(k+i)!}\left[\mu_{uk} - \frac{(1+I_{uk})a^2}{(k+i+1)^24\bar{\sigma}^2}\right].$$

Then, $\mathbb{E}\{p_{ukt}(H_{ukt})|H_{uk(L_0-L)}\}$ for $t \in \{0,\ldots,L-1\}$ can be calculated by (B.9), and substituted into (B.1) to calculate the average power consumption during the next slow time-scale time slot.

## B.2   Calculation of Average User Throughputs

Average throughput of the $u$th user during the next slow time-scale time slot over the $k$th subcarrier is given by

$$R_{uk} = \mathbb{E}\left\{\frac{1}{L}\sum_{t=0}^{L-1}r_{ukt}(H_{ukt})\Big|H_{uk(L_0-L)}\right\}$$

$$= \frac{1}{L}\sum_{t=0}^{L-1}\mathbb{E}\{r_{ukt}(H_{ukt})|H_{uk(L_0-L)}\}, \qquad\qquad (\mathrm{B.10})$$

where $H_{uk(L_0-L)}$ is the CSI transmitted to the cloud at the $L_0$th fast time-scale time slot within the current slow time-scale time slot. Since $Pr(H_{ukt}|H_{uk(L_0-L)})$ is given by (5.6) with $\bar{\sigma}^2 = (1-\rho^{2(t+L-L_0)})\sigma_{uk}^2/2$ and $s = \rho^{(t+L-L_0)}\sqrt{H_{uk(L_0-L)}}$, from (5.6), (5.7) and (5.8),

$$\mathbb{E}\{r_{ukt}(H_{ukt})|H_{uk(L_0-L)}\} = \int\limits_0^\infty \Delta f \log_2\left(1+\frac{y\left[\mu_{uk}-\frac{1+I_{uk}}{y}\right]^+}{1+I_{uk}}\right)Pr(H_{ukt}=y|H_{uk(L_0-L)})dy$$

$$= \int\limits_{(1+I_{uk})/\mu_{uk}}^\infty \Delta f \log_2\left(\frac{y\mu_{uk}}{1+I_{uk}}\right)Pr(H_{ukt}=y|H_{uk(L_0-L)})dy$$

$$= \int\limits_{(1+I_{uk})/\mu_{uk}}^\infty \frac{\Delta f}{2\bar{\sigma}^2\ln(2)}\ln\left(\frac{y\mu_{uk}}{1+I_{uk}}\right)e^{-(s^2+y)/2\bar{\sigma}^2}$$

$$\sum_{k=0}^\infty \frac{y^k(s/2\bar{\sigma}^2)^{2k}}{(k!)^2}dy.$$

(B.11)

By substituting $x^2 = y/\bar{\sigma}^2$, (B.11) can be written as

$$\mathbb{E}\{r_{ukt}(H_{ukt})|H_{uk(L_0-L)}\} = \frac{2\Delta f}{\ln(2)}\int\limits_b^\infty x\ln\left(\frac{x}{b}\right)e^{-(x^2+a^2)/2}I_0(ax)dx,$$ (B.12)

where $a = s/\bar{\sigma}$, $b = \sqrt{(1+I_{uk})/(\bar{\sigma}^2\mu_{uk})}$, and $I_0(\theta)$ is given by (B.4).
Furthermore, from (B.5),

$$\int xe^{-(x^2+a^2)/2}I_0(\alpha x)dx = -Q_1(\alpha, x).$$ (B.13)

Then, from (B.12) and (B.13),

$$\mathbb{E}\{r_{ukt}(H_{ukt})|H_{uk(L_0-L)}\} = \frac{2\Delta f}{\ln(2)}\int\limits_b^\infty \ln\left(\frac{x}{b}\right)\frac{d(-Q_1(a,x))}{dx}dx$$

$$= \frac{2\Delta f}{\ln(2)}\int\limits_b^\infty \frac{1}{x}Q_1(a,x)dx$$

$$= \frac{2\Delta f}{\ln(2)}\int\limits_b^\infty \frac{1}{x}e^{-(x^2+a^2)/2}\sum_{k=0}^\infty\left(\frac{a}{x}\right)^k I_k(ax)dx$$

$$= \frac{2\Delta f}{\ln(2)}\sum_{k=0}^\infty \frac{a^{2k}}{k!2^k}\int\limits_b^\infty \frac{1}{x}e^{-(x^2+a^2)/2}dx$$

$$+ \frac{2\Delta f}{\ln(2)}\sum_{k=0}^\infty \frac{a^{2k}}{2^k}\int\limits_b^\infty \frac{1}{x}e^{-(x^2+a^2)/2}\sum_{l=1}^\infty \frac{(ax/2)^{2l}}{l!(l+k)!}dx.$$

(B.14)

By following an integration process similar to the integration of the second term on the right hand side of (B.6), (B.14) can be simplified as follows.

$$
\mathbb{E}\left\{ r_{ukt}(H_{ukt}) \big| H_{uk(L_0-L)} \right\}
$$

$$
= \frac{\Delta f}{\ln(2)} \left[ \sum_{k=0}^{\infty} \frac{a^{2k}}{k! 2^k} e^{-a^2/2} \mathrm{E}_1(b^2/2) \right.
$$

$$
\left. + e^{-(a^2+b^2)/2} \sum_{i=0}^{\infty} \sum_{k=0}^{\infty} \sum_{l=0}^{\infty} \frac{a^{2(i+k+1)}}{2^{i+k+1}} \frac{(i+l)!(ab/2)^{2l}}{l!(i+l+1)!(i+k+l+1)!} \right]
$$

$$
= \frac{\Delta f}{\ln(2)} \left[ \mathrm{E}_1(b^2/2) \right.
$$

$$
\left. + e^{-(a^2+b^2)/2} \sum_{i=0}^{\infty} \sum_{k=0}^{\infty} \sum_{l=0}^{\infty} \frac{a^{2(i+k+1)}}{2^{i+k+1}} \frac{(i+l)!(ab/2)^{2l}}{l!(i+l+1)!(i+k+l+1)!} \right]
$$

$$
= \frac{\Delta f}{\ln(2)} \left[ \mathrm{E}_1(b^2/2) + e^{-(a^2+b^2)/2} \sum_{i=0}^{\infty} \sum_{k=0}^{\infty} \sum_{l=0}^{\infty} \frac{(a^2/2)^{(i+k+1)}(ab/2)^{2l}}{l!(i+k+l+1)!(i+l+1)} \right].
$$

$$
\tag{B.15}
$$

By calculating $\mathbb{E}\{r_{ukt}(H_{ukt})|H_{uk(L_0-L)}\}$ for $t \in \{0,\ldots,L-1\}$ using (B.15), the average throughput over the next slow time-scale time slot can be calculated using (B.10).

## B.3  Calculation of Average Interference

Normalized average interference to the $u$th user's communications over the $k$th subcarrier during the next slow time-scale time slot is given by

$$
I_{uk} = \mathbb{E}\left\{ \frac{1}{L} \sum_{t=0}^{L-1} I_{ukt} \right\}
$$

$$
= \frac{1}{L} \sum_{t=0}^{L-1} \sum_{v \in \mathcal{U}^{(c)}\setminus u} \mathbb{E}\{I_{ukt}^{(v)} | H_{vk(L_0-L)}^{(u)}, H_{vk(L_0-L)}\},
$$

$$
\tag{B.16}
$$

where $I_{ukt}^{(v)}$ is the normalized interference introduced by the $v$th user to the $u$th user over the $k$th subcarrier during the $t$th time slot; and $H_{vk(L_0-L)}^{(u)}$ is the normalized power gain of the channel between the $v$th user and the BS to which the $u$th user is connected to, during the $L_0$th fast time-scale time slot within the current slow time-scale time slot. As channels are time-correlated,

$$\mathbb{E}\{I_{ukt}^{(v)}|H_{vk(L_0-L)}^{(u)}, H_{vk(L_0-L)}\}$$

$$= \int\limits_0^\infty \int\limits_0^\infty x\Big[\mu_{vk} - \frac{1+I_{vk}}{y}\Big]^+ Pr(H_{vkt}^{(u)} = x|H_{vk(L_0-L)}^{(u)})Pr(H_{vkt} = y|H_{vk(L_0-L)})dxdy$$

$$= \int\limits_0^\infty \int\limits_{(1+I_{vk})/\mu_{vk}}^\infty x\Big(\mu_{vk} - \frac{1+I_{vk}}{y}\Big)Pr(H_{vkt}^{(u)} = x|H_{vk(L_0-L)}^{(u)})Pr(H_{vkt} = y|H_{vk(L_0-L)})dydx.$$

$$(B.17)$$

Since different wireless channels fade independently,

$$\mathbb{E}\{I_{ukt}^{(v)}|H_{vk(L_0-L)}^{(u)}, H_{vk(L_0-L)}\}$$

$$= \int\limits_0^\infty xPr(H_{vkt}^{(u)} = x|H_{vk(L_0-L)}^{(u)})dx \int\limits_{(1+I_{vk})/\mu_{vk}}^\infty \Big(\mu_{vk} - \frac{1+I_{vk}}{y}\Big)Pr(H_{vkt} = y|H_{vk(L_0-L)})dy.$$

$$(B.18)$$

By substituting (5.6) and (B.2) into (B.18),

$$\mathbb{E}\{I_{ukt}^{(v)}|H_{vk(L_0-L)}^{(u)}, H_{vk(L_0-L)}\}$$

$$= \mathbb{E}\{H_{vkt}^{(u)}|H_{vk(L_0-L)}^{(u)}\}\mathbb{E}\{p_{vkt}(H_{vkt})|H_{vk(L_0-L)}\}$$

$$= (2\bar{\sigma}^2 + s^2)\mathbb{E}\{p_{vkt}(H_{vkt})|H_{vk(L_0-L)}\}$$

$$= \big[(1 - \rho^{2(t+L-L_0)})(\sigma_{vk}^{(u)})^2 + \rho^{2(t+L-L_0)}H_{vk(L_0-L)}^{(u)}\big]\mathbb{E}\{p_{vkt}(H_{vkt})|H_{vk(L_0-L)}\},$$

$$(B.19)$$

where $(\sigma_{vk}^{(u)})^2$ is the average normalized power gain of the channel over the $k$th subcarrier between the $v$th user and the BS to which the $u$th user is connected to. Since $\mathbb{E}\{p_{vkt}(H_{vkt})|H_{vk(L_0-L)}\}$ is given by (B.9), $I_{uk}$ can be calculated by substituting (B.19) into (B.16).

# References

1. Cisco Visual Networking Index: Global Mobile Data Traffic Forecast Update, 2016–2021 [Online], Feb 2017. Available: http://www.cisco.com
2. Millimeter Wave Propagation: Spectrum Management Implications, Federal Communications Commission, Technical Report Bulletin Number 70, July 1997
3. Qiao, J., Shen, X., Mark, J., Shen, Q., He, Y., Lei, L.: Enabling device-to-device communications in millimeter-wave 5G cellular networks. IEEE Commun. Mag. **53**(1), 209–215 (2015)
4. Tse, D., Viswanath, P.: Fundamentals of Wireless Communication. Cambridge University Press, Cambridge (2005)
5. Zhou, Y., Zhuang, W.: Throughput analysis of cooperative communication in wireless ad hoc networks with frequency reuse. IEEE Trans. Wirel. Commun. **14**(1), 205–218 (2015)
6. Cheng, H.T., Zhuang, W.: Joint power-frequency-time resource allocation in clustered wireless mesh networks. IEEE Netw. **22**(1), 45–51 (2008)
7. Lei, L., Kuang, Y., Shen, X., Lin, C., Zhong, Z.: Resource control in network assisted device-to-device communications: solutions and challenges. IEEE Commun. Mag. **52**(6), 108–117 (2014)
8. Hwang, I., Song, B., Soliman, S.: A holistic view on hyper-dense heterogeneous and small cell networks. IEEE Commun. Mag. **51**(6), 20–27 (2013)
9. Ferrus, R., Sallent, O., Agusti, R.: Interworking in heterogeneous wireless networks: comprehensive framework and future trends. IEEE Wirel. Commun. **17**(2), 22–31 (2010)
10. Ismail, M., Zhuang, W.: A distributed multi-service resource allocation algorithm in heterogeneous wireless access medium. IEEE J. Sel. Areas Commun. **30**(2), 425–432 (2012)
11. Pei, X., Jiang, T., Qu, D., Zhu, G., Liu, J.: Radio-resource management and access-control mechanism based on a novel economic model in heterogeneous wireless networks. IEEE Trans. Veh. Technol. **59**(6), 3047–3056 (2010)
12. IEEE Standard for Information Technology - Telecommunications and Information Exchange Between Systems Local and Metropolitan Area Networks - Specific Requirements Part 11: Wireless LAN Medium Access Control (MAC) and Physical Layer (PHY) Specifications. IEEE P802.11-REVmb/D12, pp. 1–2910 (2012)
13. 3GPP: LTE; Evolved Universal Terrestrial Radio Access (E-UTRA) and Evolved Universal Terrestrial Radio Access Network (E-UTRAN); Overall Description; Stage 2. Technical Report. TS 36.300 V11.6.0 (2013)
14. Liang, H., Zhuang, W.: Cooperative data dissemination via roadside WLANs. IEEE Commun. Mag. **50**(4), 68–74 (2012)

© The Author(s) 2018

A.T. Gamage, X.(S.) Shen, *Resource Management for Heterogeneous Wireless Networks*, SpringerBriefs in Electrical and Computer Engineering, DOI 10.1007/978-3-319-64268-0

15. Liu, P., Tao, Z., Narayanan, S., Korakis, T., Panwar, S.: Coopmac: a cooperative MAC for wireless LANs. IEEE J. Sel. Areas Commun. **25**(2), 340–354 (2007)
16. Neely, M., Modiano, E., ping Li, C.: Fairness and optimal stochastic control for heterogeneous networks. IEEE/ACM Trans. Netw. **16**(2), 396–409 (2008)
17. Chang, H.S., Fard, P., Marcus, S., Shayman, M.: Multitime scale Markov decision processes. IEEE Trans. Autom. Control **48**(6), 976–987 (2003)
18. Fodor, G., Dahlman, E., Mildh, G., Parkvall, S., Reider, N., Miklos, G., Turanyi, Z.: Design aspects of network assisted device-to-device communications. IEEE Commun. Mag. **50**(3), 170–177 (2012)
19. Doppler, K., Rinne, M., Wijting, C., Ribeiro, C., Hugl, K.: Device-to-device communication as an underlay to LTE-advanced networks. IEEE Commun. Mag. **47**(12), 42–49 (2009)
20. Yu, C.-H., Doppler, K., Ribeiro, C., Tirkkonen, O.: Resource sharing optimization for device-to-device communication underlaying cellular networks. IEEE Trans. Wirel. Commun. **10**(8), 2752–2763 (2011)
21. Lei, L., Zhong, Z., Lin, C., Shen, X.: Operator controlled device-to-device communications in LTE-advanced networks. IEEE Wirel. Commun. **19**(3), 96–104 (2012)
22. Li, P., Guo, S., Miyazaki, T., Zhuang, W.: Fine-grained resource allocation for cooperative device-to-device communication in cellular networks. IEEE Wirel. Commun. **21**(5), 35–40 (2014)
23. Yu, C.-H., Tirkkonen, O., Doppler, K., Ribeiro, C.: Power optimization of device-to-device communication underlaying cellular communication. In: IEEE International Conference on Communications (ICC), 2009, pp. 1–5
24. Song, W., Zhuang, W.: Packet assignment under resource constraints with D2D communications. IEEE Netw. **30**(5), 54–60 (2016)
25. Wu, Y., Chen, J., Qian, L.P., Huang, J., Shen, X.S.: Energy-aware cooperative traffic offloading via device-to-device cooperations: an analytical approach. IEEE Trans. Mob. Comput. **16**(1), 97–114 (2017)
26. The 1000x Data Challenge [Online], June 2014. Available: http://www.qualcomm.com/solutions/wireless-networks/technologies/1000x-data
27. Xu, J., Wang, J., Zhu, Y., Yang, Y., Zheng, X., Wang, S., Liu, L., Horneman, K., Teng, Y.: Cooperative distributed optimization for the hyper-dense small cell deployment. IEEE Commun. Mag. **52**(5), 61–67 (2014)
28. Lu, S.-H., Wang, L.-C., Chiang, T.-T., Chou, C.-H.: Cooperative hierarchical cellular systems in LTE-a networks. IEEE Syst. J. **9**(3), 766–774 (2015)
29. Liang, Y., Goldsmith, A., Foschini, G., Valenzuela, R., Chizhik, D.: Evolution of base stations in cellular networks: denser deployment versus coordination. In: IEEE International Conference on Communications (ICC), May 2008, pp. 4128–4132
30. Zheng, J., Cai, Y., Liu, Y., Xu, Y., Duan, B., Shen, X.: Optimal power allocation and user scheduling in multicell networks: base station cooperation using a game-theoretic approach. IEEE Trans. Wirel. Commun. **13**(12), 6928–6942 (2014)
31. Gamage, A.T., Shen, Q., Shen, X.: Cloud assisted resource management for hyper-dense small cell networks. In: 2015 IEEE Global Communications Conference (GLOBECOM), Dec 2015, pp. 1–6
32. Zhu, J., Fapojuwo, A.: A new call admission control method for providing desired throughput and delay performance in IEEE802.11e wireless LANs. IEEE Trans. Wirel. Commun. **6**(2), 701–709 (2007)
33. Bianchi, G.: Performance analysis of the IEEE 802.11 distributed coordination function. IEEE J. Sel. Areas Commun. **18**(3), 535–547 (2000)
34. Gamage, A.T., Shen, S.: Uplink resource allocation for interworking of WLAN and OFDMA-Based femtocell systems. In: IEEE International Conference on Communications (ICC), Budapest (2013), pp. 4664–4668
35. Song, W., Zhuang, W.: Multi-service load sharing for resource management in the cellular/wlan integrated network. IEEE Trans. Wirel. Commun. **8**(2), 725–735 (2009)

36. Xu, J., Jiang, Y., Perkis, A.: Multi-service load balancing in a heterogeneous network. In: Wireless Telecommunications Symposium (WTS), Apr 2011, pp. 1–6

37. Song, W., Zhuang, W., Cheng, Y.: Load balancing for cellular/wlan integrated networks. IEEE Netw. **21**(1), 27–33 (2007)

38. Nasser, N., Hasswa, A., Hassanein, H.: Handoffs in fourth generation heterogeneous networks. IEEE Commun. Mag. **44**(10), 96–103 (2006)

39. Taha, A.-E., Hassanein, H., Mouftah, H.: On robust allocation policies in wireless heterogeneous networks. In: First International Conference on Quality of Service in Heterogeneous Wired/Wireless Networks, Oct 2004, pp. 198–205

40. Hasswa, A., Nasser, N., Hassanein, H.: Generic vertical handoff decision function for heterogeneous wireless. In: Second IFIP International Conference on Wireless and Optical Communications Networks, March 2005, pp. 239–243

41. Luo, C., Ji, H., Li, Y.: Utility-based multi-service bandwidth allocation in the 4G heterogeneous wireless access networks. In: IEEE Wireless Communications and Networking Conference (WCNC) (2009), pp. 1–5

42. Niyato, D., Hossain, E.: A noncooperative game-theoretic framework for radio resource management in 4G heterogeneous wireless access networks. IEEE Trans. Mob. Comput. **7**(3), 332–345 (2008)

43. Xue, P., Gong, P., Park, J.H., Park, D., Kim, D.K.: Radio resource management with proportional rate constraint in the heterogeneous networks. IEEE Trans. Wirel. Commun. **11**(3), 1066–1075 (2012)

44. Wang, P., Jiang, H., Zhuang, W.: Capacity improvement and analysis for voice/data traffic over WLANs. IEEE Trans. Wirel. Commun. **6**(4), 1530–1541 (2007)

45. Wang, H.-S., Moayeri, N.: Finite-state Markov channel-a useful model for radio communication channels. IEEE Trans. Veh. Technol. **44**(1), 163–171 (1995)

46. Ismail, M., Gamage, A., Zhuang, W., Shen, X., Serpedin, E., Qaraqe, K.: Uplink decentralized joint bandwidth and power allocation for energy-efficient operation in a heterogeneous wireless medium. IEEE Trans. Commun. **63**(4), 1483–1495 (2015)

47. Altman, E.: Constrained Markov Decision Processes. Chapman and Hall/CRC, Boca Raton (1999)

48. Puterman, M.: Markov Decision Processes: Discrete Stochastic Dynamic Programming. Wiley, New York (1994)

49. Wong, C.Y., Cheng, R., Lataief, K., Murch, R.: Multiuser OFDM with adaptive subcarrier, bit, and power allocation. IEEE J. Sel. Areas Commun. **17**(10), 1747–1758 (1999)

50. Boyd, S.P., Vandenberghe, L.: Convex Optimization. Cambridge University Press, Cambridge (2004)

51. Bertsekas, D.P.: Non-linear Programming. Athena Scientific, Belmont (2003)

52. Alam, M.S., Mark, J.W., Shen, X.S.: Relay selection and resource allocation for multi-user cooperative OFDMA networks. IEEE Trans. Wirel. Commun. **12**(5), 2193–2205 (2013)

53. Tharaperiya Gamage, A., Alam, M.S., Shen, S., Mark, J.: Joint relay, subcarrier and power allocation for OFDMA-Based femtocell networks. In: IEEE Wireless Communications and Networking Conference (WCNC), Shanghai, Apr 2013, pp. 679–684

54. Kelley, C.T.: Solving Nonlinear Equations with Newton's Method. Society for Industrial and Applied Mathematics, Philadelphia (2003)

55. Abramowitz, M., Stegun, I.A.: Handbook of Mathematical Function with Formulas, Graphs, and Mathematical Tables. Applied Mathematics Series. NBS, vol. 55. Dover, New York (1964)

56. WINNER II Interim Channel Models, EC FP6, Technical Report D1.1.1 v1.0, Dec 2006. [Online]. Available: http://www.ist-winner.org/deliverables.html

57. Rappaport, T.: Wireless Communications, 2nd edn. Prentice Hall, Upper Saddle River (2002)

58. Shen, H., Cai, L., Shen, X.: Performance analysis of TFRC over wireless link with truncated link-level ARQ. IEEE Trans. Wirel. Commun. **5**(6), 1479–1487 (2006)

59. Yu, C.-H., Doppler, K., Ribeiro, C., Tirkkonen, O.: Resource sharing optimization for device-to-device communication underlaying cellular networks. IEEE Trans. Wirel. Commun. **10**(8), 2752–2763 (2011)

60. Feng, D., Lu, L., Y. Yuan-Wu, Li, G., Feng, G., Li, S.: Device-to-device communications underlaying cellular networks. IEEE Trans. Wirel. Commun. **61**(8), 3541–3551 (2013)

61. Wang, J., Zhu, D., Zhao, C., Li, J., Lei, M.: Resource sharing of underlaying device-to-device and uplink cellular communications. IEEE Commun. Lett. **17**(6), 1148–1151 (2013)

62. Xu, C., Song, L., Han, Z., Zhao, Q., Wang, X., Cheng, X., Jiao, B.: Efficiency resource allocation for device-to-device underlay communication systems: a reverse iterative combinatorial auction based approach. IEEE J. Sel. Areas Commun. **31**(9), 348–358 (2013)

63. Xu, C., Song, L., Han, Z.: Resource Management for Device-to-Device Underlay Communication. Springer, Berlin (2014)

64. Gamage, A.P.K.T., Rajatheva, N.: A simple equalization technique to minimize ICI in OFDMA-based femtocell networks. In: 2011 IEEE 22nd International Symposium on Personal, Indoor and Mobile Radio Communications, Sept 2011, pp. 112–116

65. Gamage, A.T., Liang, H., Zhang, R., Shen, X.: Device-to-device communication underlaying converged heterogeneous networks. IEEE Wirel. Commun. **21**(6), 98–107 (2014)

66. Gamage, A., Rajatheva, N., Codreanu, M.: Resource allocation for OFDMA-based relay assisted two-tier femtocell networks. In: International Symposium on Wireless Communication Systems (ISWCS), 2011, pp. 834–838

67. Zhang, N., Cheng, N., Gamage, A., Zhang, K., Mark, J., Shen, X.: Cloud assisted HetNets toward 5G wireless networks. IEEE Commun. Mag. **53**(6), 59–65 (2015)

68. Marotta, M., Kaminski, N., Gomez-Miguelez, I., Zambenedetti Granville, L., Rochol, J., Dasilva, L., Both, C.: Resource sharing in heterogeneous cloud radio access networks. IEEE Wirel. Commun., **22**(3), 74–82 (2015)

69. Shen, Q., Liang, X., Shen, X., Lin, X., Luo, H.: Exploiting geo-distributed clouds for a e-health monitoring system with minimum service delay and privacy preservation. IEEE J. Biomed. Health Inform. **18**(2), 430–439 (2014)

70. Stolyar, A., Viswanathan, H.: Self-organizing dynamic fractional frequency reuse in OFDMA systems. In: IEEE Conference on Computer Communications (INFOCOM), Apr 2008, pp. 13–18

71. Mahmud, A., Hamdi, K.: A unified framework for the analysis of fractional frequency reuse techniques. IEEE Trans. Commun. **62**(10), 3692–3705 (2014)

72. Ding, M., D. Lopez-Perez, Xue, R., Vasilakos, A., Chen, W.: Small cell dynamic TDD transmissions in heterogeneous networks. In: IEEE International Conference on Communications (ICC), June 2014, pp. 4881–4887

73. Deb, S., Monogioudis, P., Miernik, J., Seymour, J.: Algorithms for enhanced inter-cell interference coordination (eICIC) in LTE HetNets. IEEE/ACM Trans. Netw. **22**(1), 137–150 (2014)

74. Balachandran, K., Kang, J., Karakayali, K., Rege, K.: Network-centric cooperation schemes for uplink interference management in cellular networks. Bell Labs Tech. J. **18**(2), 23–36 (2013)

75. Huang, H., Trivellato, M., Hottinen, A., Shafi, M., Smith, P., Valenzuela, R.: Increasing downlink cellular throughput with limited network MIMO coordination. IEEE Trans. Wirel. Commun. **8**(6), 2983–2989 (2009)

76. Foschini, G., Karakayali, K., Valenzuela, R.: Coordinating multiple antenna cellular networks to achieve enormous spectral efficiency. IEE Proc. Commun. **153**(4), 548–555 (2006)

77. Guler, B., Yener, A.: Uplink interference management for coexisting MIMO femtocell and macrocell networks: an interference alignment approach. IEEE Trans. Wirel. Commun. **13**(4), 2246–2257 (2014)

78. Lee, H.-H., Park, K.-H., Ko, Y.-C., Alouini, M.-S.: Codebook-based interference alignment for uplink MIMO interference channels. J. Commun. Netw. **16**(1), 18–25, Feb. 2014.

79. Samarakoon, S., Bennis, M., Saad, W., Latva-Aho, M.: Backhaul-aware interference management in the uplink of wireless small cell networks. IEEE Trans. Wirel. Commun. **12**(11), 5813–5825 (2013)

80. Sharma, N., Badheka, D., Anpalagan, A.: Multiobjective subchannel and power allocation in interference-limited two-tier OFDMA femtocell networks. IEEE Syst. J. **PP**(99), 1–12 (2014)

81. Biton, E., Cohen, A., Reina, G., Gurewitz, O.: Distributed inter-cell interference mitigation via joint scheduling and power control under noise rise constraints. IEEE Trans. Wirel. Commun. **13**(6), 3464–3477 (2014)
82. Cai, Y., Yu, F., Bu, S.: Cloud radio access networks (C-RAN) in mobile cloud computing systems. In: IEEE Computer Communications Conference Workshops (INFOCOM WKSHPS), Apr 2014, pp. 369–374
83. Qureshi, A.: Power-demand routing in massive geo-distributed systems. Ph.D. dissertation, Massachusetts Institute of Technology (2010)
84. Liang, G., Kozat, U.: Fast cloud: Pushing the envelope on delay performance of cloud storage with coding. IEEE/ACM Trans. Netw. **22**(6), 2012–2025 (2014)
85. Proakis, J.: Digital Communications, 4th edn. McGraw-Hill, New York (2001)
86. Gradshteyn, I., Ryzhik, I.: Tables of Integrals, Series and Products, 7th edn. Academic, New York (2007)
87. Hoffman, H.: A method for globally minimizing concave functions over convex sets. Math. Program. **20**(1), 22–32 (1981)
88. Gamage, A., Liang, H., Shen, X.: Two time-scale cross-layer scheduling for cellular/WLAN interworking. IEEE Trans. Commun. **62**(8), 2773–2789 (2014)

Printed in the United States
by Bookmasters

Printed in the United States
By Bookmasters